Where We Agree: Report of the National Coal Policy Project

About the Book and Editor

Where We Agree:
Report of the National Coal Policy Project
Volume 1
edited by Francis X. Murray

The report of the National Coal Policy Project is entitled *Where We Agree* to focus attention on the broad areas of agreement reached by the project participants. This one-year project brought together leading individuals from environmental and industry groups to seek consensus on important national policy issues related to the use of coal in an environmentally and economically acceptable manner. Task forces were responsible for exploring specific areas of interest: transportation, air pollution, fuel utilization and conservation, energy pricing, and emission charges.

Volume 1 contains the full task force reports on transportation, air pollution, fuel utilization and conservation, energy pricing, and emission charges. Because of its length and technical nature, the mining task force report makes up all of Volume 2. The mining report defines--on a regional and national basis--the technical, social, legal, and administrative requirements for ensuring that coal is mined in an environmentally sound manner. A summary of the project, available from the Center for Strategic and International Studies (CSIS), provides a synthesis of the project's findings and individual summaries of each task force report.

Consensus was reached on more than 150 recommendations related to such issues as locations and patterns of coal development, coal reserve quality and quantity, reclamation, the effects of development on water resources, and the socioeconomic impacts of coal development.

Francis X. Murray is director of the National Coal Policy Project, which is sponsored by the Georgetown University Center for Strategic and International Studies. Mr. Murray is director of National Energy Programs at CSIS and also serves as adjunct professor at the university's School of Foreign Service, where he lectures on energy and foreign policy. He is the author of many energy policy publications, and is currently editor of the CSIS Federal Staff Energy Seminar reports.

Project Leaders

Francis X. Quinn, S.J.
Plenary Chairman

Gerald L. Decker
Industry Caucus Chairman

Laurence I. Moss
Environmental Caucus Chairman

Project Director

Francis X. Murray

Included in This Volume

Transportation Task Force Report

Air Pollution Task Force Report

Fuel Utilization and Conservation Task Force Report

Energy Pricing Task Force Report

Ad Hoc Emission Charge Task Force Report

SPONSORED BY THE
CENTER FOR STRATEGIC AND INTERNATIONAL STUDIES
GEORGETOWN UNIVERSITY

Where We Agree: Report of the National Coal Policy Project

Volume 1

edited by Francis X. Murray

Routledge
Taylor & Francis Group
LONDON AND NEW YORK

First published 1978 by Westview Press

Published 2018 by Routledge
52 Vanderbilt Avenue, New York, NY 10017
2 Park Square, Milton Park, Abingdon, Oxon OX14 4RN

Routledge is an imprint of the Taylor & Francis Group, an informa business

Library of Congress Catalog Card Number: 78-55420

ISBN 13: 978-0-367-01769-9 (hbk)

Contents

Acknowledgements

Any undertaking of this type can only succeed because of the dedication, commitment, and hard work of those associated with it. The National Coal Policy Project was fortunate enough to have attracted a large number of highly talented individuals who gave to the project a sense of hope and worth which is so necessary. The project leaders, plenary members, task force and advisory members have all invested a piece of themselves in this project in hopes that they could collectively make a positive contribution toward solving some of our nation's more difficult energy-related problems.

Supporting this key group of participants is a small but dedicated staff which labored long and hard to keep the project going and to faithfully produce the seemingly endless drafts and redrafts and ultimately the final manuscript. A special word of thanks is in order for Cynthia Edmunds, Charles Curran, and Dale Shelly Graham, who had the difficult and thankless chore of coordinating the preparation of the final reports while still performing their regular duties. The hours were long and the work tedious and often frustrating. Jacquelyn Perry, Sheila Brown, Patricia Jacobs, Linda Farrell, and Ron Young had the unenviable job of typing the manuscripts and making the needed corrections. Finally, a special word for Dean White, Etta Pollock, and others who led us through difficult tasks such as proofreading, chart preparation, and a host of other activities. To all of the above mentioned, a thankful--WELL DONE.

Supporting this effort financially were many contributors. From industry, more than 60 companies made donations and several government agencies provided modest grants to support the project's activities. A special thanks is in order for the Ford,

Rockefeller, Sarah Scaife, and A.W. Mellon Foundations, whose belief in this concept was instrumental in getting the project off the ground without any assurances that it would be successful. In fairness to all of the supporters we should note that the recommendations contained in the report do not necessarily represent or agree with those of the sponsoring organizations.

A particular word of thanks goes to Dr. David M. Abshire, Chairman of the Center for Strategic & International Studies, without whose interest, support, and many words of wisdom this project would never have survived long enough to achieve the success which we think it represents.

Scope of the Project

The National Coal Policy Project brought together leading individuals from industry and environmental groups to seek consensus and provide guidance on the important national policy issues related to the use of coal in an environmentally and economically acceptable manner. The participants in the Project demonstrated that a process, other than an adversarial one, can be used for resolving industry and environmental differences. They found that it is possible to agree on many issues without compromising the vital interests of either side.

While the participants are members of industry and environmental groups, they took part in the Project as individuals. Therefore, the recommendations contained in Where We Agree, the report of the National Coal Policy Project, are those which the individuals believe are workable, compatible with their personal beliefs, and in the national interest.

Although only individuals from the public interest and private sectors participated, the Project clearly recognized that the Federal Government should and will be involved in the formulation of national coal policy. Most of the environmental problems associated with the use of coal present public policy choices that are often best addressed through legislation or regulation. However, the private sector has the capacity to make decisive and leading contributions if it can replace some of the conflict that has existed between industrial and environmental organizations with a constructive effort to reach policy agreements which individuals from both groups believe are desirable. The

National Coal Policy Project was designed to undertake such a task.

Unlike many other coal studies, the Project did not attempt to forecast the expected or the desired future levels of coal production or use. Considerable discussion revolved around the need for such a forecast or range of forecasts. However, the two sides agreed that while such a forecast might be interesting in an academic sense, it was more important that the ground rules or criteria for using coal be determined. Therefore, the Project sought to reach agreement on the conditions for the acceptable use of coal.

The National Coal Policy Project is also different from the many studies which have set out to develop a complete set of national policies. Instead, the Project concentrated on those areas in which the production, transportation, and use of coal have resulted in conflict between industry and environmental groups. Important questions such as mine safety, labor problems, and capital formation were not directly treated in the report because they were beyond the scope of the coal/environment interface. However, the Project recognizes that these issues should not be ignored since they have a profound effect on coal policy.

The Project did not reach agreement on all major coal policy issues that have environmental implications. For a variety of reasons, consensus was not achieved on some important issues. In some cases there was not sufficient time to address the issues adequately; in other cases the appropriate expertise or individuals were not available. In still other cases the participants agreed to disagree, recognizing that the industry and environmental members have perceptions and values that lead to different conclusions and recommendations. In cases of disagreement, they articulated their respective positions in a manner that they hope will explain their differences, with the expectation that future discussions can begin where this project left off.

ORGANIZATION OF THE PROJECT

Five task forces were established as the basic working units of the National Coal Policy Project, with each one responsible for a different area of interest: Mining, Air Pollution, Conservation and Fuel Utilization, Transportation, and Energy Pricing. Each task force was composed of an equal number of environmental and industry members. Likewise, each task force had an industry and an environmental co-chairman and vice co-chairman.

In addition to the five task forces, an Ad hoc Emission Charge Task Force was formed from members of several other task forces since discussion of emission charges overlapped their areas of responsibility.

A Plenary group, composed of all task force co-chairmen, vice co-chairmen, and the Environmental and Industry Caucus chairmen, was established to provide assistance and guidance to the task forces. The Plenary was responsible for discussing and ratifying task force recommendations, developing consensus recommendations on issues passed to it from the task forces, and assembling and approving the report of the Project.

A FINAL COMMENT

In reading the report of the National Coal Policy Project, consideration should be given not only to the individual recommendations, but to the novelty of the Project as well. The individuals on both sides came together in search of consensus on how to use this important energy resource in an environmentally sound way.

The consensus reached by the industry and environmental participants includes a number of different kinds of agreements. Some agreements are simply restatements of previously held positions which were uncovered in the course of discussion as the meanings of the language used by the two sides became clearer. Other agreements reflect positions at which the two sides arrived after having enhanced their knowledge and understanding of the range of workable solutions. Finally, some of the agreements represent negotiated positions in which one side has given ground in one area in return for gains in another. In some recommendations varying degrees of all of these factors may exist, but it is often difficult to determine which type of agreement is dominant in the resolution of a particular issue.

In any event, the large degree of success this effort has enjoyed stands in stark contrast to the paralysis of decision that has resulted from the bitter industrial-environmental disagreement over nuclear power. While the time may not be right to apply the methods of the Project to that particular dispute, nevertheless, their application should be seriously considered for the resolution of other problems for which institutional devices have not yet been developed and which need not be subjected to the rigors of the adversary process.

SECTION 1
TRANSPORTATION TASK FORCE REPORT

<u>TASK FORCE MEMBERS</u>:

J. Robert Ferguson, Jr.,
Co-Chairman

James T. B. Tripp,
Co-Chairman

W. Bruce Allen
Robert Calhoun

Mohamed T. El-Ashry
Les Moore
Patricia S. Record
Gregory A. Thomas
Gerald Westbrook

<u>Alternates</u>:
David Schwartz
William Zoller

<u>TECHNICAL STAFF</u>:

Bradley C. Gewehr

TABLE OF CONTENTS

TRANSPORTATION TASK FORCE REPORT

I. Introduction

Coal policy and transportation policy are closely interdependent. Transportation costs typically account for 40 to 50 percent of the price of coal, while coal provides the transportation sector, railroads and bargelines in particular, with an important share of traffic and revenue. In 1973, coal traffic contributed 25 percent of the tonnage carried by Class I railroads. For the 15 railroads most dedicated to coal carriage, 18 percent of their total revenues was derived from the commodity in 1975, with 58 percent of revenue coming from coal shipments in the most extreme case. Of the total tonnage carried by barge in 1974, 21 percent was coal.

The prominence of coal shipments with the transportation modes discussed herein strongly suggests that policies which bear on coal use, such as air quality standards or the mandated use of coal as a boiler fuel, will clearly have significant impacts on the transportation sector as a whole. Similarly, transportation policy will be an important determinant of the price and distribution of coal. These considerations, coupled with the perception that transportation of coal does not present a unique case calling for special treatment relative to other commodities, impelled the Task Force to formulate coal transportation recommendations within the broader context of a general transportation policy. In examining any part, the whole cannot be ignored.

The national transportation system currently suffers significant inefficiencies, in large part because of inappropriate and often conflicting policies. Significant environmental impacts are often neglected in decisions to provide transportation services and facilities, resulting in unnecessary degradation. Prices for transportation services do not necessarily reflect the full environmental and economic costs of providing those services. The result is waste of financial and physical resources.

The maintenance and expansion of the inland waterway system have been responsible for the destruction of substantial areas of wetlands. Trucks hauling coal are noisy, pollute the air, and bear much of the responsibility for the deterioration of certain public roads. Growing coal traffic may exacerbate railroad grade crossing problems. If these and other social costs are to be minimized, they must be reflected in transportation prices.

Because regulatory and public finance policies confer advantages on some modes at the expense of others, goods may often be shipped by the wrong mode. Because of unsound and inconsistent public expenditure policies, taxpayers first support the construction of unneeded facilities, and then aid carriers financially weakened by subsidized competition. Thus, while the transportation system continues to function (with the repeated infusion of federal funds), there are sound reasons for believing that substantial improvements are possible and desirable to relieve the added stress which will be generated by increased use of coal in the economy.

The growing reliance on coal energy presents the transportation system with the challenge of moving substantially increased amounts of coal. While coal already moves in large volume, changes in energy policies and consumption patterns in the United States may cause that volume to double within the next ten to fifteen years, as well as increase the average distance transported. Particularly for the railroads and the inland waterway system, these increases will severely stress the transportation system. One projection of increased coal production and demand which suggests the magnitude of future transportation needs is shown in the table, "Future Coal Production and Demand," following.

While the Transportation Task Force neither subscribes to any particular projection of coal production nor, for reasons discussed below, considers the design of a physical system to transport projected traffic within its proper scope, such projections illustrate the kinds of stress which changing energy patterns may place on the nation's transportation system.

FUTURE COAL PRODUCTION AND DEMAND[1/]

	SECTION	Prod'n. 1977 (Mill. Tons)	Prod'n. 1990 (Mill. Tons)	Mill. Tons Inc.	% Inc.	Demand 1990 (Mill. Tons)
1.	Northeast including Northern Appalachia	250	300	50	20	350
2.	Midwest and Southern Appalachia	300	450	150	50	430
3.	Central Plains and Rocky Mountains	100	280	180	180	220
4.	Southern	50	130	80	160	160
5.	Northwestern	25	40	15	60	40
	TOTAL	725	1,200	475	79	1,200

1/ Estimate by U.S. Steel Corporation. This projection is included here to illustrate the magnitude of the stress which changing energy patterns may place on the nation's transportation system. The Transportation Task Force does not subscribe to any particular projection of coal production. Nor, for reasons discussed above, does it recommend a design of a physical system to transport projected traffic. Essentially, identifying bottlenecks requires a projection of the volume and pattern of coal traffic, an engineering estimate of the condition of existing facilities, and a financial assessment of the carriers involved. Such an effort would not only require substantial resources, but would hinge on a myriad of uncertainties. Since many of these uncertainties lie at the heart of the issues of greatest importance to the Task Force, predictions of coal traffic would entail assumptions about those parameters which the Task Force is attempting to define. Because of these uncertainties, and the lack of the time and resources required to identify system bottlenecks, the Task Force chose to concentrate on policy issues. See Appendix C for a survey of some of the studies predicting the capabilities of the transportation system to meet demand projections.

Some physical bottlenecks in the system's capacity are already evident. New bridges may be needed where increased coal movement requires the railroads to lay double tracks. Equipment shortages may hamper rail service for coal and other commodities. There are potential bottlenecks on the inland waterway system as well, particularly at locks.

If carriers which require new facilities are to be capable of financing them, the correct policies to guide the choice of these additional investments should direct that, wherever market demand warrants the expansion of capacity, the required investments will be made. Conversely, physical bottlenecks can be expected to persist under policies which distort the flow of investment and cause uneconomic transportation choices to be made. Unless such policies are identified and corrected, it is likely that facilities will be constructed that promote the continued misallocation of traffic among modes, with resulting economic waste and environmental degradation.

In defining the principles and concepts which should guide and inform public policy on coal transportation, a basic performance standard is needed. Of the numerous criteria which might serve as measures of policy impacts, the Task Force concluded that *the primary goal of transportation policy should be to promote economic efficiency, with particular attention to internalizing environmental costs to ensure that they are properly weighed in transportation decisions.* The transportation policies which are deemed most desirable are those that induce the transport of coal at the lowest possible true costs.

With regard to other criteria which might serve as measures of policy impacts, safety, in the sense of accidental physical injury, was deemed of great importance. Its costs and benefits ought to be reflected in transportation prices. But transportation policy was judged to be a poor vehicle for the attainment of other social objectives, such as income redistribution. In instances when coal shipment will affect income distribution, the Task Force concluded that such impacts are best addressed directly, rather than through the narrow perspective of transportation policy. Energy efficiency and technical reliability were judged to be either subsumed under economic efficiency or to require independent treatment, not oblique treatment within transportation policy. Since the Task Force wanted to develop the best possible policy positions, political feasibility was excluded as an explicit policy criterion.

II. Policy Principles

The prime function of the transportation system should be to move commodities dependably with the fewest possible economic or environmental resources. Therefore, efficiency of resource utilization is the chief criterion for evaluating transportation policy, thereby providing a framework for the analysis of modal specific issues. This approach offers the main advantage of promoting an integrated transportation policy instead of a collection of disparate and imbalanced modal policies. The national interest lies in the provision of effective, efficient transportation through an integrated system, not in the provision of certain types of transportation, and the market will best determine how much of what types of transportation are required. Unless policies toward all modes conform to the same basic principles, some modes will be favored at the expense of others. The result would be a wasteful misallocation of traffic among modes and the misallocation of resources in the economy as a whole.

Transportation policy might be constructed from numerous perspectives. This report identifies three areas of current federal policy involvement which are particularly relevant to the central criterion of resource efficiency:

- Principles of *public finance* of transportation, with respect to both expenditures and charges;
- *Economic regulation* of coal transportation, including rate regulation and control of entry and exit; and
- *Social regulation*, including safety, environmental quality, and energy conservation considerations.

A. Public Finance

The Task Force developed its principles on the public finance of transportation in response to two basic questions. (1) Under what circumstances should there be public expenditures on transportation? (2) How should the costs of those expenditures be met?

The Task Force defined four sets of conditions under which public expenditure on transportation could be justified:

1. Where the prospective users or beneficiaries of the facility or service would pay the costs of providing it, but the market was not equipped to make the transaction, necessitating some kind of federal investment vehicle. It is agreed that, under these circumstances, the users of the facility or service (or its beneficiaries) would reimburse the government for its expenditures.

2. Where (a) a major transportation system is "*in extremis*", (b) its service is deemed essential to the public welfare, (c) the private sector will not take necessary steps for the preservation of this vital service, (d) the long-term economic viability of the system is assured so that only a one-time action is required, and (e) stringent limitations on the nature and extent of the subsidy will ensure its efficiency.

3. Where the public expenditures are exclusively for research and development of improved technology. These expenditures would be made from the general treasury.

4. Where the expenditures are used to protect environmental quality and the operation of the market would not yield a sufficiently high level of protection. In this case, the principles of cost internalization (described in the section on environmental regulation) would apply.

B. Economic Regulation

Economic regulation comprises four basic areas: the control of maximum rates, minimum rates, entry, and exit. The Task Force developed general principles of economic regulation by responding to two fundamental questions: when is regulatory intervention warranted, and what form should it take?

1. Maximum Rate Regulation: Where transportation markets are not sufficiently competitive to ensure adequate discipline in ratemaking, some maximum rate regulation by a federal agency would be warranted. A market failure could exist for particular commodities or between particular origin-destination points even if other transportation markets are workably competitive. Thus maximum rate regulation might apply to some commodities in a market and not to others, or in some markets for all commodities, depending upon the particular circumstances. Determination of the need for such regulation should be based upon a general test of competition, such as the market dominance principle of the Rail Revitalization and Regulatory Reform Act (4-R Act) of 1976 (PL 94-210). The burden of proof of the need for intervention ought to rest with the proponents of such intervention.

2. Minimum Rate Regulation: Since ratemaking below variable cost produces economic inefficiency, some federal intervention may be warranted. However, the burden of proof that a rate is below variable cost should rest with those who so contend, and not with the carrier defending the rate.2/

3. Entry: Entry should only be restricted subject to proof that aggregate shipper

2/ While there was general acceptance of this formulation, some members of the environmental caucus are skeptical of the need for any minimum rate regulation, or of the ability of the enforcing agency to discern properly where such regulation is warranted.

service would be deleteriously affected. Diversion of revenue from existing carriers is thus not necessarily inconsistent with the public interest. The minimal entry controls of assuring that a potential entrant is fit, willing, and able to serve are warranted.3/

4. Exit: Only minimal exit requirements are justifiable. No requirement for a carrier to provide a service on which it loses money can be justified. A carrier wishing to leave a market should be required to first give due notice of that intention, with some specified interval, such as 90 days between notice and actual exit. In the absence of a *prima facie* showing that currently available alternative service is inadequate to protect shipper interests, exit could be unimpeded. The burden of proof that current alternatives are inadequate should rest with those opposing the exit. In cases where alternate service can be shown to be inadequate, some limited extension of carrier service might be warranted, with the length of the extension and the service provided to be determined by the regulating agency, given some provisions for full cost recovery by the carrier.

C. Social Regulation

If the nation's transportation system used for transporting coal, as well as other commodities and products, is to be efficient, environmental costs which the system produces must be internalized, so as to become reflected in the cost of operating that mode and in the price for its services.4/ A

3/ Although consensus was reached, there is some reservation on the environmental side that any entry requirements beyond the test of fitness are justifiable.

4/ This paper does not consider the relative energy efficiencies of alternative transportation modes. If energy were efficiently priced, which it is not, an energy-efficient mode would enjoy a price advantage vis-a-vis a nonefficient mode.

reasonable technique for internalizing environmental costs for all transportation modes is to require that all modes meet all relevant federal, state, and local environmental quality standards within a set time framework. If each mode meets all relevant environmental standards, it internalizes those costs through the costs of mitigating the environmental impact. The cost of compliance is then reflected in the price of the transportation service. Full compliance with all relevant standards means that a transportation mode which must incur relatively low costs to meet all standards has a competitive advantage over another mode which would have to incur relatively high costs to meet all standards.

National, state, or local environmental quality standards should be set using a general social benefit/cost methodology.5/ Ideally, the total social benefits of achievement of a given standard should exceed the total social costs.

In thus choosing standards, the Task Force recognizes that much of the benefit of environmental protection is not susceptible to accurate measurement in dollar terms. Therefore, we must rely on the political process to intervene in these cases and make decisions which reflect the value society places on these benefits.6/

Compliance with an environmental quality standard in any particular case should not depend upon a favorable benefit/cost analysis. A lower standard might be justified for a transitional period in a particular case, only when demanding timely compliance with the national standard would be grossly disproportionate to the benefits or unjust. After the transitional period, the degrading carrier should have to comply with the national standard. This means that a variance may relax the schedule for compliance with a standard, but not relax the standard itself.

5/ An exception to this statement would appear to be congestion which is dealt with under modal specific considerations. See *infra*, III.C.3(c).

6/ The Task Force recognizes that many standards are a product of the federal bureaucracy. The problem of democratic control of the bureaucracy is not addressed in this paper, but is recognized as worthy of serious study.

If, after a limited transition period, a carrier cannot comply with a particular environmental standard with its own resources, then a local community should have the option of helping to underwrite the costs of compliance in order to avoid the loss of jobs or other adverse social impacts from the reduction or elimination of that operation.7/ The standard must still be met in order to avoid unfair treatment of those carriers complying with standards. A degraded environment should not subsidize a particular commercial operation. On the other hand, local communities should be able to establish more stringent standards through the political process.

This general approach to internalizing the costs of alternative operations which transport coal imposes uniform standards on all modes. It favors carriers which, even if their quantifiable, economic costs of operation are somewhat higher, impose lower total costs on society than do carriers which, with lower quantifiable economic costs, impose larger total social costs. This result is consistent with the economic efficiency criterion.

However, this process of internalizing costs by requiring compliance with standards is no better than the standards themselves. As a general matter, national environmental quality standards (e.g., water, air, noise, point source effluent and emission controls) have been defined and are reasonably satisfactory for many parameters of concern. More important, standards for "nonpoint" sources of environmental degradation are typically non-existent or may be inadequate and thus are typically not internalized. Moreover, general standards dealing with such complex environmental problems as groundwater depletion, congestion, nonpoint source runoff and wetlands protection may not be comparable or equally enforceable among the different modes. Such disparities would lead to inefficiencies just as disparities in economic regulation and public investment criteria do.

7/ Similarly, local communities may subsidize local branch line operations. See, infra, note 10.

The principal environmental problems for each of the four modes are as follows:8/

Mode	Environmental Problem
1. Inland Waterways	Wetlands Water quality Channelization Sediment and flow disequilibrium Anadromous fish losses
2. Railroads	Noise Congestion Air quality Land use impacts Disruption of wild-life
3. Pipelines	Groundwater and surface water depletion Water quality Land and water course impacts
4. Trucks	Air quality Noise Highway congestion Land use

These problems and general standards for evaluating and controlling them are discussed in the mode-by-mode analysis, following.

8/ Although high-voltage transmission lines are clearly functional alternatives to the four transportation modes, the Task Force did not give transmission lines explicit consideration. The chief environmental problems associated with this "transportation" strategy involve mine-mouth generation, a matter more germane to the Mining and Air Pollution Task Forces. A Situation Review of high-voltage transmission lines is included in Appendix D.

III. Mode-Specific Recommendations

The Transportation Task Force selected and discussed on a mode-by-mode basis those issues which it perceived: (1) are directly relevant to the transportation of coal; (2) are of current national importance; and (3) pertain to the economic and environmental efficiency of the coal transportation system.

A. Rail

1. Public Finance

a. The rail subsidies set forth in Title V of the 4-R Act, 9/ while satisfying the Task Force criteria of being visible, of attempting to secure repayment of public funds by beneficiaries, and of being directed toward

---

9/ Title V of the 4-R Act implements two types of rail subsidies. The first plan requires the Secretary of Transportation to administrate a Rail Fund, which is to be used for the purchase of specially issued railroad stock (the redeemable preference share). Under the second mechanism, federal funds are used to guarantee repayment of obligations negotiated by railroads. In the event that a railroad defaults on an obligation, "the full faith and credit" of the U.S. government guarantees that payment of interest and principal will be made; to the Secretary of Transportation falls the task of recovering whatever assets the railroad has.

These subsidy mechanisms are alternative means of accomplishing the same objective. Redeemable preference shares cost the railroads considerably less as a means of finance than loans negotiated in the private market and backed by the treasury. Neither seems to be a particularly good bargain for the federal government, however.

potentially profitable segments of the rail network, were nonetheless judged a poor use of public funds because they are unaccompanied by redress of the more basic afflictions of the rail system, and thus their long-term viability could not be assured. Policies corequisite to these subsidies should include:

- rationalization of rail service pricing;

- elimination of subsidies to competing modes; and

- relaxation of exit restrictions.

It was further noted that inasmuch as subsidy repayment schemes require expensive administration, provide little assurance of repayment (and in the case of redeemable preference shares, do not provide for the repayment of the full public cost), and offer incentives to be continued indefinitely through a series of vague financial instruments, a direct cash subsidy would be a preferable mechanism in many cases.

b. Under the Task Force criteria, there is little justification for the federal subsidization of local, low-density branch lines (as performed under Title VIII of the 4-R Act). Moreover, where the purpose of a subsidy is to protect the interests of shippers located along a noneconomic rail line, the least costly and most efficient subsidy would be one paid directly to the shippers and limited in size and duration to the costs of a one-time transition from rail to an alternative mode.10/ In sum, local service subsidies are an inappropriate use of federal funds.

10/ If a branch line operation is to be subsidized, such decisions should be made and the costs borne by the state or other local jurisdiction.

c. Tax subsidies (not included in the 4-R Act) are a poor idea, meeting none of the Task Force criteria for transportation subsidies; they are invisible, tend to be perpetual, are exchanged for no particular service, and require no repayment by beneficiaries. The Task Force concluded that under its own established public finance principles, no new subsidies of the rail mode appear to be warranted.

2. Rail Economic Regulation

a. In general, railroads should be allowed flexibility in establishing rates. The only objective of maximum rail rate regulation should be to prevent the exploitation of market power by carriers. In some rail transport markets, rates are restrained by intermodal and/or intramodal competition, the need of the carrier to retain the shipper's markets, and shipper flexibility in location, production, and marketing. Where such discipline is inadequate and maximum rate regulation is retained, it must be accompanied by evidence of market power with the burden of proof resting upon the proponents of intervention and regulation. *The regulatory agency may be such a proponent.*

b. Minimum rate regulation should consist of a protest against a rate believed to be below variable costs. The protest should include the reason that the rate is believed to be harmful. If a protestant presents such a *prima facie* case, the ICC will then set the case and assume the role of investigator with the burden of proof that a rate is below variable cost and is harmful to rest upon the protestant.

c. Restrictions on entry should be minimized under the general supposition that entry is beneficial unless demonstrated otherwise by a party able to prove that shipper service will be impeded. A regulatory proceeding to determine only questions of impairment of service and general fitness to serve should be retained as a precondition to entry.

d. No carrier should be required to continue a service which loses money. Hence, exit from a market should not be impeded (given a specified period between notice of the intent to exit and actual exit) where the ICC determines that transportation alternatives available to shippers are adequate. If they are not adequate some extension of service or a transition subsidy to shippers may be warranted.

e. Railroads should be permitted to engage in contract ratemaking. (The Task Force did not agree on whether or not contract rates should be made public.)

3. Rail Social Regulation

The most significant railroad environmental problems appear to be noise, congestion and air quality impacts of unit train shipments. All of these impacts could become more severe if coal unit train shipments increase significantly. However, air quality and noise impacts, can, in our opinion, be managed readily and effectively under existing federal statutes.

a. Air quality impacts should not be significant since the establishment and enforcement of technology-based or other emission control air quality standards for diesel engines should be accomplished without difficulty under the 1977 Amendments to the Clean Air Act of 1970.

b. Regulation of railroad noise is now supposed to occur largely at the federal level. Pursuant to Section 17 of the Noise Control Act of 1972, 42 U.S.C. Sec. 4901, 4916, EPA is required to promulgate noise emissions standards:

> "...setting such limits on noise emissions resulting from operation of the equipment and facilities of surface carriers engaged in interstate commerce by railroad which reflect the degree of noise reduction achievable through the application of the best available technology, taking into account the cost of compliance." Section 17(a)(1) of the Noise Control Act.

The federal noise standards preempt any local regulation except where, as Section 17(c)(2) of the Noise Control Act provides, EPA and DOT determine that the local standard "is necessitated by special local conditions and is not in conflict with" EPA regulations.11/

We support the emphasis on national noise standards, particularly for moving equipment with provision for local variances necessitated by special conditions. However, since the noise standards are supposed to reflect both the noise level of an individual train movement and the number of train operations, any EPA noise standard for railroads should reflect anticipated increases

---

11/ Originally, EPA established noise emission standards limited to interstate railcars and locomotives. 40 C.F.R. Sec. 201.11, 201.12. In a court challenge brought by the Association of American Railroads, a federal appeals court has recently ruled that the Noise Control Act requires EPA to promulgate nationwide standards for all rail "equipment and facilities" in order to avoid conflicting noise control regulations in every jurisdiction along rail routes and to assure the maximum practical uniformity in regulating the noise characteristics of interstate rail carriers. *Association of American Railroads et al., v. Costle*, __ F 2d __, No. 76-1353 (D.C. Cir. Aug. 23, 1977). Thus, EPA is now obliged to issue noise emission standards for all rail equipment, mobile and nonmobile.

in unit train operations in particular regions of the country. As a general objective, total noise exposure from increasing unit train shipments should not rise as measured by monitoring stations located in logical places. Increases in train operations must be appropriately compensated for through reductions in noise levels per unit train shipment. Since unit trains customarily use continuously welded track (due to the reduced maintenance costs which result) which reduces rail noise emissions considerably, we would anticipate that unit trains would have no difficulty meeting any reasonable EPA noise standards. In communities which have acute railroad noise problems already, sound directing devices in populated areas are technologically and economically feasible to abate noise pollution.12/

c. There are many potential problems associated with rail traffic congestion. The increased train traffic in the West threatens to cut many towns in half during significant portions of the

---

12/ EPA has promulgated a standard 90 decibels for new moving locomotives at a distance of 100 feet to be met by December 31, 1975. Since the EPA National Standard is based on certain assumptions about the frequency of rail operations nationwide, the environmentalists believe the standard should be reduced somewhat if the number of unit train movements increases significantly from 1976 levels. On the other hand, the industrialists are of the opinion that because of the intermittent nature of railroad noise there would be no mitigation of health or nuisance values by further noise reduction despite increased frequency. The detailed EPA Railroad Noise Emission Standards appear at 40 C.F.R. Part 201. For additional information on EPA's methodology and background data, see Background Document for Railroad Noise Emission Standards (Dec. 1975) EPA-550/9-76-005.

day, and to disrupt livestock and wildlife grazing patterns and some farming operations. We recognize that on the low side little increase in coal transportation may occur nationally. On the other hand, the National Energy Plan projected increases over the 1976 level of annual national coal production by roughly 70 percent in 1985. Others have made projections on coal transportation from the Northern Plain states to the Midwest and East which show increases from 1976 levels of 250-500 percent. One-hundred unit trains, each hauling 10,000 tons of coal, could be required to operate in an easterly direction each day under these projections. With this level of traffic, some routes will carry some 25-33 trains per day in each direction. Operating at 10-20 MPH through towns,13/ unit trains would thus occupy trackage space an average of 10-30 minutes per hour during a 24-hour period.

Furthermore, as traffic congestion increases, accidents and fatalities may be expected to rise proportionately unless adequate steps are taken to control congestion.14/

This conflict between the town and the railroads, although now raised to new heights, has been litigated and fought over as long as the railroads have existed. A large amount of precedent concerning the duties and obligations of each party has been laid

13/ In towns with grade crossings, train speed may be as low as seven MPH, compounding congestion problems.

14/ Estimated annual fatalities for unit train shipments from Wyoming to a reference 1000 MW plant in the Midwest is 1.45. Dvorak, *et al.*, *The Environmental Effects of Using Coal for Generating Electricity*. (Prepared for the U.S. Nuclear Regulatory Commission by Argonne National Laboratory, NuReg-0252, 1977.)

down by federal and state courts. Because of the complexities of developing a congestion standard applicable to unit train shipments, it is useful to see how these doctrines should be applied in this new situation.

According to American case law, it is well established that it is "the duty of the railroad company whose road intersects a public street to construct and maintain without compensation a crossing separated from the street by a bridge or otherwise when reasonable public safety and necessity are required."15/

15/ 65 Am. Jur. 2d Railroads Section 276 (1972), citing at n. 78 City of St. Paul v. Great Northern Ry. Co., 138 Minn. 25, 163 N.W. 788 (1917). See, 74 C.J.S. Railroads Section 151 (1951) citing at n. 29 Boston and M.R.R. v. Worcester County Comm'rs.,15 N.E. 2nd 455 (1938) (railroad should pay, but only if bridge is necessary.) See also 65 Am. Jur. 2d Railroads Section 267 (1972), "When a railroad operates upon public streets, it assumes the burden of sharing on a fair and reasonable basis the cost of any public improvement needed to further public safety and convenience made necessary by the growth of the community," citing Atcheson, T.&S.F.R. Co. v. Public Utilities Commission 346 U.S. 346 (1953). Similarly, the granting to the railroad of a right of way across a highway does not give it the right to obstruct the highway and exclude the public use of it. See 65 Am. Jur. 2d Railroads Section 269 (1972), citing at n. 38 Denver v. Denver and R.G.R. Co., 63 Colo. 574 (1917). See also Pennsylvania Utilities Commission v. Borough Souderton, 231A.2d 875 (1967) in which a Pennsylvania court recognized that "(t)raffic congestion is an element to be considered in determining the public interest in cases involving highway-rail crossings." 231 A. 2d at 878.

Although no doubt such regulation does affect interstate commerce, the states and towns through their police power have been allowed to resolve these conflicts.16/ There is no apparent present need to withdraw this

---

16/ 65 Am. Jur. 2d Railroads Section 274 (1972). See Nashville C. & St. L.R. Co. v. White, 278 U.S. 456 (1929). When the railroads have challenged the constitutionality of the police power of the state and town to impose the entire cost of constructing crossings and installing signal devices, the state's and town's actions have been sustained by federal and state courts. See R.R. Erie Railroad Co. v. Bd. Public Utilities Commissioners,254 U.S. 394 (1921); City of Gainesville, Ga. v. Southern Ry. Co., 447 F. 2d 562 (1971); Atcheson T. & S.F.R.Co. v. Public Utilities Commission, 547 P. 2d 234 (1976). In many western states, under present state laws, the decisions of the allocation of costs between railroads and towns have usually been left up to Public Utility Commissions or similar state agencies. Still, in some particular cases, the railroads have private contractual arrangements with the municipalities. In Utah, the Public Utility Commission regulates this aspect, Utah Code Ann. Section 54-4-15 (1975), as it also does in Montana, Mont. Rev. Codes Ann. Section 72-101. In Wyoming, the costs are to be apportioned "in accordance with justice." Wyo. State Section 37-64. Schemes for the apportionment of costs when a bridge or underpass must be built are provided by statute in Montana and Wyoming. Pursuant to Mont. Rev. Codes Ann. Section 72-707 (1975 Supp.), the Public Service Commission, after considering the safety, necessity, and convenience to the traveling public of building the structure, has authority to impose the entire cost on the railroad or to apportion the costs. Under Wyo. Stat. Section 37-235.2, the Public Service Commission decides the priority for safety devices and grade crossings after considering the increased rail traffic and traffic volume. Although under 37-235.3(a) the railroad can only have a 33 percent burden maximum of the costs of the safety devices, significantly, no such limit is set with (Cont'd.)

power from the states and towns. Considering the large increases in unit coal traffic forecast to go through these towns, however, it would seem to be both fair and reasonable for the railroads generally to bear the costs incurred in building necessary grade separations, installing signal devices for safety, or rerouting movements around populated areas.

In this situation, not all "costs" would be allocated to the railroads, as it might seem at first glance, since a bridge would be constructed only when "necessary" rather than at every crossing. Therefore, the town would be incurring the cost of adjusting its behavior to accommodate the new restricted access.17/ Moreover, since it is the national energy demand for more coal which may lead to this result, it seems only equitable that the brunt of these costs are placed on the railroads which in turn will be able to distribute these costs among consumers of the coal. Similar logic suggests that railroads should be primarily responsible for livestock fatalities and disruption of wildlife use patterns caused by train operations and therefore should bear the cost of mitigation measures such as

---

16/(cont'd.) respect to the apportionment of the costs of the grade separation structures under Section 37-235.3(b). Apportionment is to be "based upon the causes resulting in the need..." This would seem to indicate that, with the increased coal train traffic, the railroads should in many cases bear the entire cost.

17/ If grade separations in a community provide crossing capacity in excess of what is presently available so that the community achieves a net benefit the community should of course make a contribution. The railroads should finance construction of grade separations, etc., necessary to maintain access capacity.

construction of adequate livestock, wildlife, and farm equipment crossings.18/

Generally, the Task Force would recommend that 1975 conditions be set as the norm for evaluating unit train congestion and safety impacts until evidence is supplied that this norm is either too lax or too stringent. The railroads would therefore be required to install grade separation, line reroutings, and signaling equipment as necessary such that congestion and fatalities do not significantly increase and such that farming and grazing patterns are not detrimentally affected.

To date, we have found no study which adequately evaluates these congestion/noise/safety impacts of increased unit train movements. Considering the importance of the issues surrounding the movement of coal unit trains through towns and the difficulty of assessing the problem due to the lack of information on the subject, the

---

18/ On the issue of injury to livestock, the Supreme Court has upheld the constitutionality of state statutes that require railroads to fence rights-of-way, through lands of private persons [Minneapolis St. L. Ry. Co. v. Emmons, 149 U.S. 364 (1893)], and that imposes liability on railroad companies for livestock killed because the railroad either failed to or defectively fenced its tracks [Missouri Pac. Ry. Co. v. Humes, 115 U.S. 512 (1885)].

According to Corpus Juris Secundum (C.J.S.), in the absence of a statute or agreement providing otherwise the railroad is not obliged to fence in tracks and to provide cattleguards to protect animals (Moody v. Texas & P. Ry. Co. App. 37 So. 2d 346); (C.J.S., Sec. 177), and is only liable for livestock injured or killed due to the railroad's own negligence (C.J.S., Sec. 547, P. 1150). The railroad's liability is found by referring to the applicable state law [C.J.S., Sec. 549, Miss. Pac. R. Co. v. Hayes, 122 S.W. 2d 945 (1939)].

Transportation Task Force recommends that EPA and DOT undertake a comprehensive study immediately to determine the scope of these problems. Such a study would quantify these unit train congestion and safety impacts, estimate the costs of mitigating these impacts in accordance with our proposed norm, and propose alternative mitigation actions. This information would be useful to state agencies where local solutions are deemed appropriate.

d. A great deal of consideration was given to the environmental impacts of new rail lines since, under present circumstances, discontinuance of service on a particular line by a railroad or construction of new trackage may necessitate the preparation of an environmental impact statement pursuant to the National Environmental Policy Act. This is so because under existing ICC regulatory practice, approval of discontinuance of rail service or construction of new trackage with the introduction of new service typically constitutes a major federal action.

We are recommending substantial contraction of federal involvement in rail entry and exit decisions. Nevertheless, with the adoption of our transportation policies, entry of rail service through the construction of new rail lines may continue to constitute major federal action for NEPA purposes inasmuch as an agency finding regarding the impairment of shipper service may be required. Moreover, where a new rail line may cross federal lands or directly or secondarily affect waters of the United States and contiguous wetlands and other federal resources involving a federal permitting process, decisions regarding construction of new rail lines should properly require the preparation of an environmental impact statement.

In all these cases, the central emphasis in such environmental impact statements should be consideration of cost effective alternative routes which would minimize or avoid adverse environmental impacts on federal lands and water resources. If our economic policies are adopted, achieving regulatory parity among different transportation modes, and less regulation generally, it would be inappropriate for any responsible federal agency to discuss secondary economic impacts on other transportation modes in an environmental impact statement regarding new railroads.

B. Waterways[19/]

1. Public Finance

The inland waterway system is currently built, operated, and maintained with public funds. Because waterway users do not bear any part of these costs except through general taxation and because project evaluation and authorization procedures fail to replicate a financial test which would determine the soundness of new investment, the system has been overbuilt and is not utilized efficiently. Neither economic efficiency, nor environmental enhancement is promoted by current policies. The Task Force therefore makes the following recommendations:

a. Because of the lack of existing requisite investment institutions, public finance of the waterway system is warranted.

19/ Economic regulation of barge traffic is not discussed explicitly. For the most part, coal is carried on the inland waterway system by private or contract carriage and hence without rate regulation. For reasons stated in the general discussion of economic regulation, extension of rate, entry or exit controls to this mode would not be desirable.

b. Users and beneficiaries should pay the full costs of operating, maintaining, and repairing waterways, as well as the costs of new construction.20/ The construction costs of existing facilities are not relevant to current and future economic decisions, and may be ignored for these purposes.21/

c. Such waterway user charges should be based, insofar as possible, on the costs of operating segments of the waterway.22/ In this way, high cost, inefficient segments can be distinguished from efficient ones, and the efficiency of the system will not be impaired by cross-subsidization.

d. Where there are transitional problems caused by the location along the waterway of shippers responding to distorted price signals, phased implementation of the user charge may be warranted.

e. At the present time, the U.S. Army Corps of Engineers is responsible for implementing public investment decisions regarding operating and maintaining, and expanding and constructing the inland waterway system. In making recommendations, the Corps utilizes cost/benefit analyses. In general theory,

---

20/ The industrialists are of the opinion that imposition of waterway users charges should be coupled with strict adherence to the minimum railroad rate regulation policy noted above. See Section III.A.2(b).

21/ Robert Ferguson, Industrial Chairman of the Task Force, in view of his position as Executive Vice President of U.S. Steel Corporation, dissents from these recommendations as they relate to waterway users charges.

22/ Waterway user charges should reflect only the cost of navigation-related expenses. Costs of other water resource uses, e.g., flood control, should not be included in the charges.

any benefit/cost ratio resulting from such analyses must exceed unity before public monies may be spent. In addition to the institution of user charges, procedures utilized by the Corps of Engineers in its cost/benefit analyses should be modified as follows:

(1) In Corps of Engineers cost/benefit analyses, an attempt should be made to quantify environmental costs (subject to enumeration and definition) and to include them in the cost/benefit analysis to determine whether a new project should be authorized. Authorized projects on which construction has not been initiated or progressed substantially should be subject to reanalysis under the new rules.

(2) Continued operation of existing segments of the inland waterway system should be subject to a periodic cost/benefit analysis which includes the factors of the environmental costs of operation and maintenance procedures, barge movement and social costs and benefits.[23/] Segments for which total costs exceed total benefits should be closed down.

(3) In a situation where new construction occurs or operation of an existing segment continues, all possible cost-effective steps should be taken to avoid or to mitigate environmental damage. Where suitable, these measures should be required of the waterway users. Where it is more convenient or suitable for these measures to be taken by government or by entities other than the waterway users, the full cost should be recovered from those users through a system of user charges.

23/ The Task Force acknowledges the inherent difficulties in measuring the full costs of environmental quality factors in such analysis. The general problem is alluded to in Section II.C.

f. In the case of existing segments which do not recover operation, maintenance, and amortization costs, local communities may subsidize continued operation of that segment. The federal government should discontinue any subsidies, however.

g. To secure the most impartial financial test of the need for new facilities, projects should be privately financed to the greatest extent possible. This might be accomplished by establishing a waterway authority for each new project contemplated. The waterway authority (or a similar agency) might be empowered to issue and sell bonds or other financial instruments on the private market to raise the funds required for project construction. Such bonds should be secured solely by the revenue from the facility, and not by the "full faith and credit" of the federal government.

h. Use of projects built with private funds not secured by the U.S. Treasury may be priced at the discretion of the waterway authority or similar agency.

2. Inland Waterways--Environmental Regulation

For coal transportation purposes, the Mississippi River and its tributaries constitute the most important inland waterway. Some segments of this system, such as the Mississippi River south of St. Louis, are considered to be naturally efficient navigable waterways. Other segments of the system, such as the upper Mississippi River, can only be maintained as navigable with an elaborate system of locks and dams. Some segments, such as the Tennessee-Tombigbee Waterway now under construction, would consist in part of artificial canals and ditches with multiple locks and dams. As a consequence, the environmental impacts of maintaining, operating and constructing different segments of this waterway system differ radically.

a. Specific environmental impacts: wetlands and water quality. The canalization of streams, construction of canals, locks, dams, and wing dikes, dredging, and disposal of dredged spoils in connection with construction and maintenance of inland navigation waterways typically adversely affect levels of dissolved oxygen and turbidity, nutrient levels, anadromous fisheries, sediment flows, and surface and groundwater hydrology, and contribute to incremental downstream or backwater flooding or sedimentation.

The construction and maintenance of some segments of the inland waterways system also typically destroy fresh or coastal wetlands.24/ The destruction of these wetlands can adversely affect water chemistry and, in turn, the quality of its capability to sustain biological activity. Federal navigation projects may also have adverse secondary impacts on wetlands and water quality through increased urban runoff resulting from the intensified development of areas near the waterways. Some segments of the inland waterway system, such as the main stem of the lower Mississippi River, can be maintained with relatively less adverse impact on wetlands and water quality than other less efficient segments of the system which can be maintained or expanded for navigation only with severe adverse environmental impacts.

24/ See, R.M. Darnell, Impacts of Construction Activities in Wetlands of the United States (1976), prepared by Tereco Corp. for EPA (no. 600/3-76-045): EPA, Office of Water Planning & Stds., Nonpoint Source Pollution Guidance Hydrologic Modifications (1976). See also, Arthur D. Little, Inc., Report on Channel Modifications, pp. 195-269 (1973) (submitted to Council on Env'tl. Quality). For a quantified analysis of the lower Mississippi River bend cutoff channelization program on downstream flow lines, see, R.E. Louque, "New Flowline for the Lower Mississippi River," 12 Water Resources Bulletin, No. 5 (October, 1976).

With respect to inland waterways, EPA and EPA-approved state standards for water quality and wetland protection are presently in force. The Corps of Engineers, however, which designs, constructs and operates the inland waterway system, has broad variance rights under the 1972 Federal Water Pollution Control Act Amendments and its own regulations issued under Section 404, 33 U.S. C. Sec. 1344.[25/]

The legal and *de facto* power of the Corps to disregard environmental standards should not exceed that of the other modes; i.e., the fact that it is a public agency should not give waterway construction or operation any advantage over railroad or pipeline construction or operation with respect to compliance with environmental standards.

b. Recommendations for natural resource preservation goals. First, although the Corps may maintain that it is not legally required to comply with water quality standards in connection with maintenance of existing segments of the inland waterway system,[26/] it should have to comply with those standards with respect to the construction of new waterways, and maintenance of new or expanded waterways.

Second, the Corps should be required to manage the inland waterways in a manner which complies with water quality standards and FWPCA water management plans in connection with all maintenance dredging operations and barge operations by July, 1983. In evaluating the impact of its construction and dredge and fill operations on water quality, and determining compliance with water quality standards, the Corps should be required

---

25/ 33 CFR Sec. 209.145, as amended, 33 CFR, Parts 320 through 329.

26/ Minnesota v. Hoffman, 543 F. 2nd 1198 (8th Cir. 1976), Cert. denied, Sub Nom, Minn. v. Alexander, 45 U.S.L.W. 3703 (U.S. April 26, 1977). See also, Sec. 61 and 67 (b)(R) of the Clean Water Act of 1977, amended Sec. 313, 401 and 404 of the 1972 Federal Water Pollution Control Act Amend., 33 U.S.C. 1323, 1341 and 1344.

to consider not only construction-related impacts, but also secondary impacts which arise from barge operations, induced development and destruction of wetlands, and major alterations of riverine ecosystems, i.e., degradation of aquatic biological systems. The Corps does not conduct such analyses at present.

Third, the Transportation Task Force agrees that further destruction of wetlands is detrimental to the ecology as a whole, and possibly to the economy as well. As a policy goal, the Corps' activities should not result in any net destruction of wetlands or any net reduction in wetland values based on the U.S. Fish and Wildlife Service wetland value system Habitat Evaluation Procedures, in connection with new or ongoing construction and expansion of waterways as of the present time and maintenance dredging operations in existing waterways as of July 1983. Some flexibility may be needed to avoid adverse impacts with respect to other possible social costs which may occur as a result of strict compliance with this recommendation. Where it can be demonstrated that the best available technology and management techniques, including consideration of the total costs of their application in a particular case, will not assure strict compliance with the foregoing wetlands policy, a variance, including an additional transition period before full compliance, may be appropriate. The best practicable disposal technologies for dredged spoil, developed through continued agressive research programs of the sort conducted by the U.S. Corps of Engineers' Waterways Experiment Station at Vicksburg, Mississippi, should make it feasible for the Corps of Engineers to comply with that standard.27/

27/ See, for example, Fourth Annual Report, Dredged Material Research Program, U.S. Corps of Engineers' Waterways Experiment Station, Vicksburg, Mississippi (January, 1977).

In certain segments of the inland waterways system, turbidity, caused by a wide range of point source and agricultural runoff and transportation-related nonpoint source pollution, may be a major water quality problem. Continuing high levels of turbidity in such rivers and backwater areas may degrade biological resources. In some of these segments, resuspension of particulate matter caused by barge traffic and bank erosion caused by barge wash may be significant sources of continuing high levels of turbidity. Studies to identify sources of turbidity and means of its control resulting from barge operation and nonpoint sources should be undertaken by EPA and the U.S. Fish and Wildlife Service. These studies should also evaluate both abatement and management techniques, including the use of marginal green belts along the waterways, controls on barge velocities, and draft and horsepower. If appropriate, such techniques should be incorporated in Corps of Engineers and U.S. Coast Guard regulations relating to the operation of the affected segments of the waterway system.

C. Coal Slurry Pipelines

1. Economic Regulation

a. Regulatory Framework

Historically, pipelines[28/] have been regulated as common carriers,[29/] despite service characteristics which are inherently private or contractual due to the

28/ Natural gas pipelines which are regulated to serve a variety of objectives not related to transportation policy.

29/ "(T)ransportation of oil or other commodity ...by means of pilelines...shall be considered and held to be common carriers within the meaning of (the Interstate Commerce Act)." 49 U.S.C. Sec. 1, 1(b).

limited flexibility of the mode. This regulatory practice has more to do with policy toward the oil industry than with transportation policy. In order to prevent unfair pipeline rates, Congress included provisions in the Hepburn Act of 1906 giving the pipelines common carrier status and directing the ICC to regulate pipeline rates.

Coal slurry pipelines, as common carriers by classification under the Interstate Commerce Act (ICA) are obliged, at least in theory, to:

--provide transportation upon reasonable request thereof (Sec. 1 (4)); 30/

--establish reasonable rates with other such carriers and file tariffs, subject to suspension and rate imposition by the Commission (Sec. 1(4); 15(1), (3) and (7));

--establish just and reasonable rates, fares, charges and classifications (Sec. 1(4), (5) and (6));

--avoid practices discriminatory to any shipper (Sec. 3);

--file reports with the Commission (Sec. 20(1));

--adopt a uniform system of accounts (Sec. 20(3), (4) and (5)).

While these obligations are explicit in the Interstate Commerce Act, precise and detailed application of any of these provisions within the coal slurry pipeline context remains somewhat theoretical in the absence of any judicial interpretations of the subject.

30/ All section citations refer to the Interstate Commerce Act.

Although common carriers by classification, the coal slurry pipelines are contract carriers by their nature. They tend to be relatively inflexible in terms of both routing and the commodities which they are physically capable of carrying. These considerations, coupled with extreme capital intensity, make indispensable the use of relatively long-term contracts between the coal shipper and the slurry pipeline operator.

Pipelines are also unique within the common carrier classification in that they are not subject to entry or exit supervision nor are they subject to the "commodities clause," a limitation applicable to other common carriers which prevents the carrier from transporting goods in which it has an ownership interest. Notably, the several bills which have been introduced in Congress to confer eminent domain powers upon coal slurry proprietors would impose entry restrictions similar to those applicable to other common carriers and would apply the "commodities clause" restriction.

By contrast, railroads are not common carriers by statutory definition, and yet are more pervasively regulated than are the slurry pipelines. The Interstate Commerce Commission has determined carriage by contract to be an unfair and discriminatory practice by railroads.31/ Moreover, the Act applies entry and exit restrictions and carrier-owned commodity limitations with respect to this mode. This regulatory asymmetry between the potentially competing rail and pipeline modes raises serious policy questions relevant to the charter of the Transportation Task Force.

31/ Apparently, *de facto* contracts have been sanctioned to some extent with respect to carriage by unit trains.

b. The right of eminent domain

The railroad industry contends that, if the coal slurry lines are allowed to operate, they will capture the long-haul high-volume, highly profitable shipments, in effect "skimming the cream," leaving the less profitable shipments to the railroads.32/ Such competition with the slurry pipelines, the railroads contend, would decrease their overall ability to make the capital investment necessary to provide transportation services for coal shippers.33/ Railroads are a flexible transportation system that can carry many commodities while the slurry lines are more rigid and capture the few isolated large lucrative shipments without providing the array of beneficial services that railroads do.

The slurry pipeline's response is that the railroads will not be capable of handling all of the increased traffic that the nation's need for coal will demand.34/ Almost certainly, the introduction of coal slurry lines in the West

---

32/ In 1974, coal accounted for 21.5 percent of the Burlington Northern's total freight by weight, which comprised 11.1 percent of the company's total revenue. This was the railroad's single largest source of income. They contend that the loss of these and anticipated future revenues will result in reduction in the railroad's ability to finance a long-term commitment, leading to a further loss of jobs, and causing increases in the prices of transporting other commodities. Furthermore, in making their present investment decisions, the railroads further contend that they have been depending on these anticipated revenues.

33/ See,A. Sargent, "Western Coal Transportation: Unit Trains or Slurry Pipelines," prepared for U.S. DOT Report No. DOT-TST-26T (1976) pp. 76-78. See also, Sargent, supra, at p. 10 quoting Moody, Trans. Manual 1975. (NY: Moody Invest. Serv., Inc., 1975), p. 362.

34/ Pipeline proponents refer to the Nat'l Academy of Engineering Reports conclusion that to handle a demand of 1.26 billion tons of coal by 1985, it will take four new 25-million-ton slurry lines, 60 new
(cont'd. on next page)

will take away some future revenues for which the railroads were hoping. However, there is nothing inherently wrong with these normal consequences of increased competition provided that each mode is on an equal footing in the market.

Therefore, the Task Force concluded that *pipelines should be granted the right of eminent domain, contingent upon the removal of regulatory policies which prohibit railroads from making long-term contracts.* Since slurry pipelines are by nature contract carriers, impartial treatment of the railroads requires that they be given the right to negotiate contract rates as well; hence, the Task Force reiterates a position taken in the earlier discussion of railroad regulation. Granting railroads contract rate-making powers should be a condition of granting federal eminent domain powers to the coal slurry pipelines. The dispute between the railroads and the pipelines would then be settled within the market setting.

c. Entry and exit

Again perceiving no cause for exception to its general principles of economic regulation, the Task Force concluded that *entry restrictions on slurry pipelines are unwarranted beyond a basic test of fitness to serve.* It is sometimes argued that entry restrictions are necessary to prevent destructive competition, which is said to result from a situation in which two carriers serve a market that could be most efficiently served by one. A "natural monopoly" is

---

34/ *(cont'd.)* east rail-barge systems, 70 new west rail-barge systems, 2 new gas pipelines, 8,000 new locomotives, 150,000 new 100-ton hoppers. See, "The Battle Over Coal Slurry," Wall Street Journal (John Ryan) Dec. 8, 1975. Cf. "Transporting the Nation's Coal--A Preliminary Assessment," Report to the Sec'y. of Trans. from the Coal Transportation Task Force (Jan. 1978), U.S. Dept. of Transportation.

said to exist in such a case.[35] In the event that a "natural monopoly" does exist, it should serve as a sufficient deterrent to entry by other firms, obviating the need for policy attention.[36]

The Task Force also observed that no valid reason exists for barring railroads from entry into slurry pipeline ventures. Railroads ought to be able to pursue least-cost solutions to their coal transportation needs.

Consistent with the principle that no service ought to be provided which cannot generate the revenue to cover its own costs, *the Task Force dismissed the notion of restricting exit of slurry pipelines*. Given the contractual agreements between shippers and pipelines, the Task Force saw no problems likely to emerge in this area.

35/ A coal slurry pipeline, like any pipeline, may have certain natural monopoly features because of economies of scale associated with pipelines. On the other hand, there may be certain diseconomies of scale associated with the problems of obtaining water and sufficient coal in a particular location. In addition, there may be technological diseconomies of scale. For these reasons, the Task Force did not consider that any natural monopolistic features which may be associated with a pipeline warranted any fundamental change in our basic entry principles.

36/ If the risks of entry were borne within the private sector, it would be quite unlikely that a slurry pipeline could be financed unless there was enough potential traffic to make it profitable. Since current traffic would be contracted, there is little chance of traffic diversion from the existing pipeline to the new entrant; new traffic would have to enter the picture, and presumably this traffic would be in excess of the capacity of the existing line.

d. Rate regulation

Neither minimum nor maximum rate regulation is appropriate to coal slurry pipelines. Since the opportunity for cross-subsidization is not present, there is no reason for minimum rates to be regulated. Maximum rate regulation is not warranted either, since pipeline rates will be negotiated and contracted. An upper ceiling will be imposed on rates by, among other factors, the competitiveness of the market for coal and the threat of rail or other competition with the pipeline.

2. Social Regulation of Coal Slurry Pipelines

With respect to coal slurry pipelines, the major environmental impacts relate to ground and surface water depletion through consumptive loss or export of water, and degradation of water quality. Land use impacts may not be significant so long as construction activities do not violate water quality and wetlands protection standards.37/

a. Water quality impacts

Water quality impacts arise at the point where the slurry is discharged. This will occur both at the end of the pipeline and along the route of the

---

37/ For instance, in establishing the 100-foot right-of-way needed for the proposed 1036-mile Wyoming-Arkansas pipeline, over 12,000 acres of land will be used (U.S. Congress, House Comm. on Interior and Insular Affairs, Coal Slurry Pipelines Legis. Hrgs. testimony of Cong. Mark Andres (ND), p. 114. However, the amount of land permanently occupied would be 840 acres. According to the Slurry Transport Assn. (STA), however, it takes only about 4 days to construct a line from the time the ditch is open to the time it is filled in so that restoration of land can begin, "How a Coal Slurry Pipeline Protects the Environment and Community Life," by STA). For a discussion of land use impacts of pipelines and other modal corridors, see Section E, supra.

pipeline when dumping of the slurry is needed to free a clogged section of the line. This slurry water may contain significant concentrations of pollutants, particularly dissolved solids, as well as suspended solids.[38] Point source pollution standards for such discharges exist. However, specific effluent standards and limitations for toxic pollutants, where they exist at all, are just now in a state of development. The end of pipeline discharge can be eliminated if the water is reused for utility cooling purposes.[39] The water quality discharge problem is then converted into a utility air quality, wastewater discharge or solid waste problem. So long as the utility meets all standards for air and water quality and solid waste management, water quality costs will be internalized. Similarly, so long as the slurry dumped along the lines is pumped back into the pipeline, it will not cause water quality problems. Water quality in streams may also be jeopardized when pipelines cross rivers, either from construction effects or from breaks in the slurry line. Both problems can be rendered negligible by requiring the best engineering practices.

b. Water depletion impacts

The water depletion problem is far more complex. It arises from the fact that withdrawal of water from the aquifer or surface water systems may lead to both reductions in the groundwater table and direct loss of instream values. This "overdraft" occurs when the amount of water extracted exceeds recharge from

---

38/ Coal Slurry Pipeline Hearings, supra, letter from Russell Train, then Administrator of the EPA, to the House Committee, p. 46.

39/ Energy Transport Systems, Inc., "Slurry Pipelines, Innovation in Energy Transportation" (October, 1975) p. 19.

natural and imported sources.[40/] This can cause harmful economic and environmental short- and long-term impacts.[41/] In the short run, a lowering of groundwater levels leads to increase in the pumping lifts required.[42/] This, in turn, leads to increased costs to other water users and increased energy consumption with concomitant air quality degradation.[43/] There also may be long-term harmful environmental impacts such as subsidence and aquifer compaction.[44/] When certain types of formations are overdrafted, subsidence and compaction can lead to the storage capacity of the aquifer being irreversibly decreased. Furthermore, since some groundwater and surface waters are hydrologically connected, significant reductions in a water table may cause stream flow reductions with attendant adverse consequences to recreation uses, water quality, and stream biota.[45/]

40/ Willey, W.R. (EDF staff scientist) "Groundwater in the California Quandary," prepared for the 11th Biennial Conference on Groundwater, Fresno, Cal. (Sept. 15, 1977) p. 2, describes some of the economic and environmental implications of groundwater overdrafts. In Cal., the major cause of this problem is withdrawal of water for agricultural irrigation. Each additional demand for groundwater can intensify a potential overdraft problem in the absence of a carefully designed and embraced management plan.

41/ In its 1972 report, the Nat'l. Water Comm. concluded that the depletion of groundwater aquifers at rates exceeding recharge was one of the "three principal problems of groundwater loss." "Water Policies for the Future," (Final Report of the Nat'l. Water Commission, June 1973) p. 232.

42/ Willey, *supra*, note 15, "In 1977, such overdrafts (in CA) electricity pumping costs will...fall in the $25-$65 million range." pp. 2-4.

43/ *Ibid.*, p. 4.

44/ *Ibid.*, pp. 5-7;R. Train, *supra* note 38.

45/ U.S. Interior, Water for Energy Mgmt. Team Northern Great Plains Area, p. II-14 (Jan. 1975); U.S. Interior, Water Mgmt Team, Upper CO River Basin, pp. 35, 46; R. Train, *supra* note 38.

Coal slurry lines, which are expected to obtain their water mostly from groundwater sources, need about one ton of water to move one ton of coal.46/ The proposed slurry lines, some of which plan to transport over 25 million tons of coal a year, will need large amounts of water. Therefore, without adequate groundwater management, the introduction of coal slurry pipelines in the West could aggravate depletion problems.47/

The Task Force members believe that reform of water management programs is needed.48/ The prior appropriation system with its "first in time, first in right" doctrine encourages waste, and does not adequately consider the environmental implications of water usage.49/ Only recently

---

46/ "Coal Slurry Pipleline...," 53 North Dakota L. Rev. (1977); See also, "An Analysis of Technical and Legal Issues Raised by the Development of Coal Slurry Pipelines," 13 Houston L. Rev. (1976) pp. 449, 456-458, 528.

47/ According to one source, there have already been problems with drawdowns in wells and reduction in stream flows in the area of the Black Mesa coal slurry pipeline. Telephone conversation with A. Harris, U.S. Dept. of Interior, Land and Water Resources Div. (July 25, 1977). Data from U.S. Geological Survey fails to confirm this assertion, but concedes that additional information is needed. (2nd Int'l. Conf. on Slurry Technology--Slurry Transport Assn, Mar. '77).

48/ "Modernizing State Water Rights Law; Some Suggestions for New Directions," 1974 Utah L. Rev. 760 (1974).

49/ Among the states that use the Prior Appropriations Doctrine are CO (CO Const'n. Art. XVI, Sec. 6 and Art. XVII, Sec. 5 and 6); MT (MT Rev. Code Ann. 89-885, R.C.M. 1947); NV (NV Rev. Stat. Sec. 533.370) ND (ND Century Code 61-04-06); WY (WY Stat. Sec. 41-203).

have some of the states started to expand the traditional concepts of "beneficial use" and "injury to existing users" to consider environmental factors such as instream uses and recreation.[50] The many proposed energy developments in the West, such as the coal slurry pipelines, would place new demands on scarce western water resources.[51] In the case of slurry pipelines, a full-scale water recycling system would effectively internalize groundwater depletion costs, but estimates of the cost of such a system make this solution unlikely.

The states' instinctive efforts to protect their water through legislation such as anti-export statutes do not represent long-term solutions to the problem.[52] Moreover, the constitutionality of these state statutes restricting export of water from the

---

50/ NV Rev. Stat. Sec. 533.030 (recognizing recreation as beneficial use); CO Rev. Stat. Sec. 148-121-317 (Supp. 1973) ("to preserve the natural environment to a reasonable degree.") The State of Idaho upheld stream rights in Southern Idaho Fish and Game Ass'n. et al., v. Picabo Live Stock, Inc., 528 P. 2nd 1295 (1974). In 1974 Montana passed a three-year moratorium on any further appropriations in the Yellowstone Basin in order to ensure adequate instream supplies for wildlife and aquatic habitats. MT Rev. Code Ann., 89-8-103 (1974).

51/ Water is not the only transport medium for slurried coal, however. If oil were used to carry coal in pipelines, the water problem would be avoided while conserving energy transportation corridors. For instance, the projected Northern Tier crude oil pipeline could pick up slurried coal in Montana and carry it to the pipeline terminal in Minneapolis. Significant economic and environmental savings would be likely.

52/ WY Stat. 4-10.5; CO Rev. Stat. (1973), Sec. 37-81-101, et seq.; Rev. Code of MT, 1947,89-867(2) ("a use of water for slurry to export coal from MT is not a beneficial use.")

state is in serious doubt.[53] What is likely needed are institutional changes so that water management decisions can be made that reflect local, regional, and national energy, as well as environmental and other interests.

Although basic reform of western water law is desirable, it would not be realistic to condition coal slurry pipeline development on such reform. Therefore, the Task Force sought a policy with respect to coal slurry pipelines that would internalize the costs of groundwater depletion in the absence of broad water policy reform. As noted in the section on economic policy, the Task Force recommends that passage of national eminent domain legislation for coal slurry pipelines should be contingent on the granting of contract rights to railroads. In addition, to protect against groundwater depletion, the Task Froce recommends that specific coal slurry pipelines be granted the privilege of eminent domain only after a federal administrative finding that operation of such a pipeline will not cause or contribute to incremental net depletion [54] of any aquifer, or damage any instream uses, including fish and wildlife needs during low flow periods.[55]

53/ After the case City of Altus, OK v. Carr, 255 F. Supp. 828, affd. (W.P. Texas, 1966) per curiam 385 U.S. 35 (1966) where the Court held that water was an article of commerce, these state statutes may well be found unconstitutional. See, "It's Our Water," 10 Land and Water Law Review 119 (1975).

54/ Industrial member Gerald Westbrook dissents in favor of a policy prohibiting "permanent damage" of aquifers. Environmental members prefer the textual formulation due to lack of confidence that "permanent damage' can be adequately ascertained in advance of its occurrence.

55/ Some exception to this policy might be acceptable if costs of avoiding stream damage are large in comparison to benefits foregone and reasonable alternatives are not available.

D. Trucks[56/]

1. Public Finance

The increasing use of trucks to haul coal in some areas of the Appalachian mining region has caused the rapid deterioration of public roads. Trucks are the preferred mode in areas where mining operations do not generate enough coal traffic to warrant the construction of a private infrastructure or where coal markets are near mines. The continuous movement of heavy trucks along roads that are often not designed to bear such traffic accelerates road damage which, according to some studies increases as a cube of the increase in truck weight. As a result, roads subject to this coal traffic are frequently both very expensive to maintain and require more public funds than are available for the purpose. The problem then is primarily to ensure that those who use the public roads to transport coal make a proper contribution to the maintenance and repair of those roads.

The current level heavy-duty truck user charges is inadequate to provide for proper road maintenance,[57/] particularly for those roads which are not a part of the Interstate Highway System. Based on studies being conducted by the Federal Highway Administration,

56/Economic regulation of truck carriage is not discussed specifically. Coal transport by truck is predominantly intrastate, thereby exempt from economic regulation by ICC. For reasons stated in the general discussion of economic regulation, control of rates, entry, or exit should be minimized for this mode.

57/ Appendix E presents a calculation of highway maintenance costs attributable to trucks. These figures are higher than those reported by other sources, e.g., Research Analysis of Factors Affecting Transportation of Coal by Rail or Pipeline, Hudson Institute, H.R. 2409-RR (April 1976). Different studies report widely varying relationships between the roadway damage caused by automobiles versus that caused by heavy trucks, but are unanimous in asserting that the trucking industry is not paying its fair share of the national highway costs.

the appropriate user charge level should be determined and properly distributed on the basis of factors affecting road wear. Such a user charge system would ideally be based on the axle weights and distance traveled by the trucks. Because of the administrative difficulties of taxing these factors directly, some combination of a tax on gross vehicle weight and fuel consumption is likely to be appropriate.

Moreover, it is not economically efficient to continue spending billions of dollars on new highway construction while deferred maintenance costs are building at a rate of several billion dollars per year. Therefore, in order to make the most efficient use of available money, the Highway Trust Fund, currently available for new construction, should be opened up to permit use of its funds for maintenance.

2. Social Regulation of Trucks

The major pollution impacts of trucks are air pollution, water pollution in the form of nonpoint source street runoff, noise, congestion and highway safety.58/ Although the Clean Air Act of 1970 prescribed the establishment of exhaust emission standards for all mobile sources, those standards failed to adequately internalize the cost of air pollution caused by trucks. The deficiency was largely cured in the 1977 Amendments to the Clean Air Act. The statutory goal for percentage emissions from truck emission standards should be at least as stringent as those for automobiles.59/ *Trucks should have to comply with standards established and*

58/ The Task Force also acknowledges that flying chunks of coal from moving trucks may pose a significant safety and env'tl. hazard. However, ameliorative strategies were not considered by the Task Force.

59/ Sec. 224 of the Act amending Sec 202(a) of the Clean Air Act of 1970, 42 U.S.C. Sec. 521, provides that EPA will set emission reg'ns. for carbon monoxide, hydrocarbons and oxides of nitrogen for new truck models, starting with model year 1979, based on the greatest degree of reduction achievable based on available technology. Truck air pollution control
(cont'd. on next page)

applicable transportation and land use control plans prepared and approved under the Clean Air Act. They should also have to comply with standards and procedures for controlling non-point source street runoff pollution established in 208 wastewater treatment management plans, prepared and approved under the 1972 Federal Water Pollution Control Act amendments. In local situations where large fleets of trucks are used to transport coal to utilities, noise and congestion problems comparable to those sometimes caused by unit trains may be created. In the case of trucks, these noise and congestion problems may arise more in eastern metropolitan areas than in small western towns. EPA has the mandate to establish noise emission regulations for motor carriers engaged in interstate commerce under Section 18 of the Noise Control Act of 1972, U.S.C. Section 4917.60/ In developing these regulations, EPA must consult with DOT "in order to assure appropriate consideration for safety and technological availability.61/ The procedures and noise level frequency relationships which EPA uses to develop these standards for motor carriers are similar to those used for rail carriers, as described above. Similar allowance for local variances also are provided.

EPA issued noise emission standards for new medium and heavy trucks (those over 10,-000 pounds) on April 13, 1976.62/ After considering the benefits and costs of truck noise level reductions, EPA has established a standard of 83 decibels at 50 feet for new trucks in an accelerating mode to take effect January 1, 1978. That standard will go down

---

59/*(cont'd.)* goals are: 75 percent reduction of NO_x by 1985 and 90 percent of CO and hydrocarbons by 1983 for heavy-duty vehicles. Sec. 224.

60/ Sec. 18(a)(3) of the Noise Control Act of 1972, 42 U.S.C. Sec. 4917 (a)(3).

61/ Sec. 18(c) of the Noise Control Act of 1972, 42 U.S.C. Sec. 4917 (c).

62/ Trans. Equipment Noise Emission Controls Medium and Heavy Trucks, 40 CFR Part 205. For add'tl. information on EPA's methodology and bkgd. data, see, Background Documents for Med. and Heavy Truck Noise Emission Reg'ns. EPA-550/9-76-008 (March 1976).

to 80 decibels in 1982. Since, accoustically, trucks contribute a substantial portion of highway noise, EPA should consider a more stringent standard for trucks by 1985.63/ Furthermore, EPA should emphasize the importance of research to make such goals technologically possible and DOT should permit local communities, which experience substantial increases in truck traffic for coal movement, to develop highway noise control plans which require trucks to meet more stringent noise standards. As a general objective, in communities which are now experiencing high levels of total truck and automobile noise exposure, overall cumulative truck noise production should not be permitted to increase from present levels.64/ Other impacts of noise and congestion may have to be resolved through local restrictions on truck traffic.

A significant increase in truck traffic associated with the transport of coal might necessitate construction of new highways. Since highways are constructed by governmental agencies which enjoy significant political power, environmental standards are frequently compromised. The major direct environmental impact of new or expanded highway construction relates to changes in land use. Probably the only technique available to internalize some of these land use costs is the NEPA process.

63/ EPA has research underway to determine what kinds of technology would be available to make trucks even quieter. It anticipates that the trucking industry may be able to meet a 75 decibel standard, using the best available technology by 1985.

64/ For general information about intrastate and interstate movements of trucks transporting coal at the present time see, Map No. 3 of the Nat'l. Energy Transportation Vol. I - Current Systems and Movements: report prepared by the U.S. Geological Survey and the Congressional Research Service at the request of the Comm. on Energy and Natural Resources and the Comm. on Commerce, Science and Transportation, U.S. Senate, Publication No. 95-15 (May 1977).

E. Modal Corridor Land Use Impacts

Railroads, pipelines, highways, and high voltage power lines utilize corridors across lands and rivers and thus can have significant impacts on land use patterns with accompanying socioeconomic effects. Significant environmental advantages may, therefore, accrue in terms of reduced land and watercourse disruptions if different modes are able to utilize the same corridor. This issue should be fully analyzed and evaluated through the NEPA process whenever invoked.

The NEPA process will be involved in cases where a grant of eminent domain power is sought in the construction of coal slurry pipeline, where federal funding is involved in the construction of a highway, where permission is sought to lay a new right-of-way across federal lands or navigable waters for a railroad or high voltage power lines, and finally, where entry for a railroad or pipeline is sought from a regulatory agency. Environmental impact statements should specifically consider the environmental advantages of reduced land use disruption in using existing modal corridors. Grants of eminent domain or rights-of-way across public lands should be provided only on the condition that the new corridors will be available for use by other modes at reasonable rates.[65] Similar considerations may be appropriate in the state permitting and environmental review processes.

Federal Regulatory agencies such as the ICC and DOT should seriously consider proposals to encourage corridor conservation and multicommodity transportation efficiences. For example, the use of petroleum pipelines in carrying slurried coal should be considered as a preferable alternative to the granting of eminent domain powers for the construction of a slurry pipeline, where technically and economically feasible.

F. Policy Recommendations for Funding of Research for All Modes

Government funding of basic scientific research that may have potential implications for the coal transportation system is clearly justified. The

---

65/ The Task Force recognizes that engineering and marketing requirements may make existing corridors incompatible with the requirements of new entrants. See also, Energy Transport Systems, Inc. v. Union Pacific Railroad Co. 435 F. Supp. 313 (P.C. WY 1977).

justification for government support of basic research is two-fold in that the risk to the individual entrepreneur that he will not be able to capture the benefits of such research is very high, and basic research results should be defused as much as possible for the benefit of all.

In general, the government should not subsidize commercial applications of developed technology within the transportation system. But in cases involving highways and waterways, where the rights-of-way are public and traditionally government-owned, there is justification for the government to support implementation of research concepts.

The Task Force points out that economic incentives for research and development work in the transportation sector have in many cases been destroyed because of inefficient and inappropriate economic policies. If these policies were modified, as we recommended, there would be less of a need for government support of applied transportation research and of the commercialization of new technologies.

There are three specific areas, based on the general principles outlined above where the federal government should support basic research:

1. *Research into cost-effective management techniques for resolving or mitigating environmental problems; for example, (1) techniques for utilizing dredge spoil material to restore wetlands and (2) government research of turbidity studies* as outlined in Section III.B.2(b).

2. *Development of additional hydrologic data regarding groundwater systems in the West which may be used as a source of water for coal slurry pipelines.* A long-term effort to inventory groundwater resources and uses would permit more accurate determinations regarding the impacts of proposed pipelines.

3. Because of the existence of serious highway maintenance problems, the government should support *research to develop better materials and methods for constructing and maintaining highways.*

In view of the size of the federal energy research budget, we would propose that water resource and pollution problems should enjoy a relatively high priority in the expenditure of these research and development funds.

IV. Impacts of Adoption of Economic and Environmental Policies on Transportation of Coal

While each of our economic and environmental policy recommendations can stand alone, they are intended as an integrated and cohesive package. Therefore, we consider it futile to attempt to assess the impact of each individually. It should also be apparent that adoption of certain policy changes with respect to one mode may require parallel action in other modes in order to secure competition, which we consider desirable. Partial adoption and implementation of our policy recommendation might benefit one mode in a manner detrimental to another mode. The following assessment is directed to the impacts of the entire group of economic and environmental policies which we propose:

A. General Impact of Policies on Transportation of Coal

The major impact of our economic and environmental policies would be the allocation of costs where they belong. Costs now defrayed by general tax revenues and costs now externalized and borne by society as a whole would be reflected in the price of coal and paid ultimately by the consumer. *Roughly speaking, general taxation would therefore be reduced to offset direct user taxation so that adoption of our economic regulatory policies will not result in any net increase in taxes in general.* For example, user charges on waterways and additional truck taxes should be offset by comparable decreases in general taxes (and a higher standard of maintenance of existing highways used for coal haulage).

The general impact of our policies would be an improvement in the efficiency of the transportation system through the elimination of economic and environmental waste. These recommendations reemphasize the general policy inherent in recent federal environmental legislation that environmental costs borne by certain segments of society should be redistributed as economic costs to users.

We would expect that our economic policy proposals would have more significant impacts on the future capacity and operational efficiency of the transportation system than our environmental policies. The in-place environmental regulations, as set forth in the 1972 Federal Water Pollution Control Act Amendments, the 1970 Clean Air Act (both as amended in 1977), the 1972 Noise Control Act, and existing federal and state administrative practices, have initiated the process of internalization of costs as transportation modes seek to comply with those standards. *In general, with the possible exception of the Task Force wetlands policy, which affects water transportation, the changes in environmental policies which we recommend are relatively minor when viewed in the context of transportation policy as a whole.*

Even the wetlands policy which we recommend will not impose significant additional costs on efficient segments of the inland waterway system--those segments which continue to be cost-justified following imposition of our user charge policy. Its successful implementation by the Corps of Engineers will require additional research and implementation of the best-available dredged spoil management technologies.

In sum, full and adequate internalization of all economic and environmental costs of all the modes should not have a significant effect on the overall costs of transporting coal. On the whole, our economic policies would be noninflationary since general taxes should be reduced by an amount comparable to additional waterway user charges or truck taxes. Indeed, total costs of transporting coal may decline, with the general taxpayer paying less to subsidize coal transportation and coal consumers paying somewhat more. *The adoption of the economic investment and regulatory policies which we are proposing, by removing the major institutional constraints, will improve the efficiency and productivity of the country's transportation system.* In this case, the capacity of the transportation system to respond to whatever demands may be put upon it by the country's energy requirements would be greatly enhanced.

The cost impacts associated with compliance with the recommended environmental standards together with adoption of our economic policies will vary from mode to mode and, therefore, will entail shifts from one mode to another. Furthermore, we would anticipate that our policy would have a significant effect on future transportation investment decisions. If adoption of our economic and environmental policies should result in significant increases in the costs of transporting coal in some instances such that other energy sources or energy conservation investment strategies would become relatively more attractive, this result is consistent with the economic efficiency criterion.

Although it is impossible to predict the quantitative results of the adoption of our economic policies on the transportation system as a whole and to individual modes specifically, we do have some empirical evidence indicating that reduced regulation of transportation systems improves the efficiency of their operation and that unregulated transportation systems operate more efficiently than regulated ones.[66]

On the other hand, if our economic and environmental policies are not adopted, then we would anticipate greater economic inefficiencies, with rising coal transport requirements, escalating taxpayer subsidies for coal transport, increased highway under-maintenance, inflationary transportation costs, inappropriate transportation investment decisions and continued environmental degradation, particularly regarding destruction of wetlands.

66/ Despite the fact that our economic policy recommendations rest more on economic logic and our own mutual judgments about the most appropriate response to existing inefficiencies in the American transportation system, we present evidence in Appendix B that adoption of the kinds of economic policies which we propose has enhanced the efficiency of operation of affected transportation systems in several instances. <u>See</u>, <u>Appendix B</u> entitled, "Empirical Examples, Where What We Propose in Regulatory Reform Has Already Been Implemented." This empirical evidence supports our overall judgment that the proposed policies will increase the efficiency of the transportation system and will specifically enhance the likelihood that the transportation system will respond in an efficient, cost-effective and environmentally sensitive way to future coal transportation demands, whatever they may be. <u>See also</u>, Appendix A.

B. Impacts of Adoption of Recommended Policies on Each Mode-Intermodal Shifts

In addition to general impacts on the transportation system adoption of our economic and environmental policy recommendations would shift movements of coal among modes. These shifts would become more evident in future years as our policies affect future investment decisions. Although precise quantitative impact analysis is not feasible, we would expect qualitative shifts as follows:

1. Impact on waterways

The Task Force waterway user charge policy would increase costs to large carriers of using all segments of the inland waterway system. Similarly, our wetlands policy would increase costs of dredging and thus costs of maintaining, expanding and constructing all segments of that system. These costs would be reflected, in turn, in segment user charges. However, these combined cost increases on efficient segments of the inland waterway system would be rather small, since they are the segments on which most of today's barged coal moves.

Almost all coal presently moves on the Lower and Upper Mississippi River, the Ohio, Cumberland, Illinois, Monongahela and Tennessee Rivers. According to an American Association of Railroads analysis, which is not apt to understate O and M costs, allocated O and M costs on the Lower Mississippi River (Cairo to Baton Rouge) are 0.13 mills per ton-mile, the Upper Mississippi (Minneapolis to Cairo) 0.45 mills per ton-mile, the Ohio, 0.41, the Cumberland, 1.04, the Illinois, 0.57, and the Monongahela, 1.40. If the average barge haul of coal is 480 miles, these added costs are small.[67] Indeed,

---

67/ It is important to note, however, that these figures reflect only operating and maintenance costs, not construction expenses which continue to be incurred on segments of the inland waterways. Future construction costs would be recouped under the user charge design which we advocate. These costs can be substantial. For examples, Barloon (Case Western
continued on next page

for the Lower Mississippi they would be only $0.20 per ton of coal shipped. Very little coal moves on the Missouri River, which is a very expensive segment (9.7 mills per ton-mile) and virtually none moves on the Arkansas. Indeed one user charge study indicates that full cost-recovery tolls will shift only some three to five percent of all waterway traffic to other modes, while closing down marginal segments at a total savings of some 50 percent of current O and M expenditures.

On the other hand, *given the adoption of our policies, the inland waterway system probably would not be expanded. Indeed, it more likely would contract, shedding marginal segments.* Certainly, highly marginal new segments now under construction, such as the Tennessee-Tombigbee or Red River Waterways, would not be built or would be closed down soon after completion. Nor would we expect an expanded Missouri River system to move coal unless high volume shipments combining rail or pipeline with water shipments should result in significantly lower user cost charges per unit of coal.

Thus, as marginal semgents are closed and future transportation capacity investments are shifted to rail and pipeline, we would expect a smaller percentage of new coal movements in the future to move by waterways than is now the case. We would also expect that our waterway economic and environmental policies would have overlapping impacts on individual segments. Both the recommended economic policy changes and environmental standards would concentrate barge tows transporting coal or other commodities on the

---

67/*(cont'd.)* Reserve University) calculates user charges on the Ohio River to recover full federal outlays at 3.0 mills (vice 0.41 mills reported herein). Moreover, the visible trend, even for O and M costs is in the direction of rather dramatic increases, unless a user charge is instituted to deter uneconomic expenditures and investments. For a discussion of the general impact of waterway user charges see Staff Study U.S. Trans. Dept Systems Center, Modal Traffic Impacts of Waterway User Charges, Rept. # SS-212-U1-32 (March 1977).

principal navigable segments of the Inland Waterway System. Much of the environmental destruction takes place on those segments of the waterway system which are the least economically efficient to operate. Thus, much of this destruction could be avoided and the environmental policies achieved by institution of an efficient system of waterway user charges.

2. Rail

Although it is difficult to estimate the cost impacts of the proposed noise and congestion policies on railroads, *The Task Force's economic policies would enhance the efficiency of the rail system and would, by reducing or eliminating subsidies to competing modes, increase real revenues.* Increased revenue, combined with more efficient utilization of rail facilities, would likely cause improved rail service quality and would allow more flexible response by the rail system to new coal shipping demands.

3. Pipeline

Because of very limited experience with coal slurry pipelines, it is meaningless to compare the impact of our policies on future pipeline costs relative to present costs. However, subject to our policies of eliminating regulatory disparities and protection of ground and surface water systems, we are supporting grants of eminent domain to coal slurry pipelines. The potential entry of pipelines would enhance competition and thus hold transportation costs to levels below those which would be expected without coal slurry pipelines transport.

4. Trucks

Heavy trucks are presently undertaxed, although they paid some $7 billion in taxes in 1975.[68/] In addition, some $4 to $5 billion in general tax revenues is appropriated for highway use in the country. If the costs

68/ Motor Vehicle Manufacturers Assn., Motor Vehicle Facts and Figures (1976).

of highway maintenance actually being absorbed were fairly internalized for heavy trucks, then the additional cost per ton of coal adjusted for taxes might be as high as $0.014/ton mile or $0.82 per ton of coal for an average haul.[69] This estimate reflects only O and M costs on existing highways and no new construction costs. In addition, many of our highways are badly undermaintained, becoming an increasingly serious problem as our highway system ages. If highways were properly maintained, these additional truck user charges would be considerably higher.

C. Intermodal Shifts

This modal cost information suggests that the proposed policies would bring about a net shift of coal movements from truck to rail and a small shift from waterway to rail. In the future, as investment decisions affect modal capacities and costs, our policies should bring about a larger shift of future coal movements from trucks and waterways to rail, and some shift from rail to pipeline.

D. Other Impacts

In addition to general modal-specific and intermodal-shift impacts, we would anticipate other implications of our transportation policies for national coal policy and, more specifically, for energy policy:

1. Generally, it is unlikely that significant amounts of western coal will be shipped to points east of the Mississippi. The low Btu per poun heat content and transportation costs will at some point balance the lower mining cost for western coal.

2. We would anticipate that the rail system in the Northeast will be upgraded and that significant amounts of eastern coal will move to New England, New York and other states in the Northeast, possibly by a combination of rail and intercoastal waterway.

69/For detailed calculations showing the derivation of this figure, see Appendix E.

3. In the West, rail transportation costs would be lower if mining were centered around rail modes or pipeline terminals, or if mining were concentrated such that new rail lines would cost less to construct and operate. Furthermore, we would expect that the location of a mine will be a factor in determining the cost of transportation and the choice of transportation mode.

4. The changes in transportation costs which the adoption of our policies would generate will not in themselves, create significant economic incentives for mine-mouth generation of electricity in the West. In addition, air and water resource considerations would militate against this tendency.

5. Our policy recommendations are likely to have some small influence on power plant siting, as the increased price of truck transport of coal (due to the internalization of road maintenance and pollution costs) will tend to make site locations near rail lines or nearer to mines relatively more attractive than truck-served sites.

E. Urgency of Policy Implementation

This report would not be complete without acknowledging that the policy recommendations herein are major, and in some respects controversial, and that the time for implementation is short if the transportation sector is to respond as well as it might to the greatly increased pressures which near-term national energy policy will exert. A sense of urgency must be conveyed, therefore, along with our recommendations. Delays in action not only perpetuate a precarious national energy profile, but can also lead to substantial cost and degradation penalties.

It is unlikely that these policy changes will be made in one legislative passage. More likely, several packages will be required over some period of time. In considering the development, promotion, implementation and ultimate administration of such new legislation, it is felt that a spirit of cooperation and facilitation is needed by politicians, administrators, environmentalists, and industrialists in order to get the necessary changes underway as rapidly as possible.

Appendix A

EMPIRICAL ESTIMATES OF THE BENEFITS AND COSTS OF ECONOMIC REGULATIONS OF SURFACE TRANSPORTATION

Recently, the Bureau of Economics of the Interstate Commerce Commission (1976) made some estimates of the benefits and costs of ICC regulation. They concluded that costs were a minus $4.8 billion (i.e., a benefit) and the benefits were $4.4 billion. Although the ICC cautions against adding the above two numbers, they do in a footnote to Table 7, and suggest that $9.2 billion is the maximum limit of attainable benefits that can be attributable to ICC regulation. However, on page 29 of the report, the ICC states that if the nonquantifiable benefits in their analysis had been quantified, "The total level of ICC annual benefits could well double that estimate," i.e., $8.8 billion. Adding the $4.4 billion to the $9.2 billion above yields a net $13.6 billion benefit attributable to the ICC per year.

This ICC study is just one of a series of studies in the last ten years assessing the benefits and costs of regulation. The Commission's is the first such study to give a quantitative statement which yields positive benefits to regulation. All other <u>quantitative</u> studies show costs of regulation. Of course, many qualitative statements both pro and con regulation have existed for many years.

Allen and Hymson (1977), for the Council on Wage and Price Stability, have refuted the ICC's above analysis. While not generating a methodology of their own, Allen and Hymson attack each constituent number of the ICC report. For instance, the largest cost of the ICC's minus $4.8 billion cost figure was a minus $3.8 billion component associated with the inefficient use of trucks. By properly applying the ICC's convoluted logic, the Commission's -$3.8 billion is changed to a +$539 million. The largest benefit of the ICC's $4.4 billion benefit was a $3.24 billion component associated with hold-downs of motor carrier and rail rates. However, using a different price index deflator than the Commission (the GNP deflator as opposed to the wholesale price index), the net benefit becomes -$1.37 billion, i.e., a cost.

Sufficient problems exist with every number in the ICC analysis and with some of the statistical methodology used in the report to discredit the Commission's analysis. For more details see Allen and Hymson (1977).

The ICC report was designed to refute Moore's (1972, 1975) study. Moore showed costs of ICC regulation in 1968 dollars of between $3.8 and $8.9 billion (or $6.5 and $15.2 in 1975 dollars - to be comparable to the ICC study). The cost comparisons between Moore and the ICC are cited in the table at the end of this appendix.

Moore's methodology is not without pitfalls as are pointed out in the ICC study. Note that the gap between the two reports is $28.8 billion, i.e., Moore's inflated to 1975 high estimate of $15.2 billion and the Commission's estimate with nonquantifiable benefits of $13.2 billion. The data are simply not available to conduct the type of analysis needed to estimate precisely the costs and benefits of regulation. The aggregate nature of the demand data with respect to geographical area as well as to commodity detail make such analysis impossible at the present time. The ICC study of benefits is fraught with difficulty of attributing benefits to some individuals. Allen and Hymson (1977) discuss the difficulties and frustrations associated with attempting to quantify ICC regulation.

Because of compromises in the data and methodology, any number presented becomes a target relatively easy to destroy given the the data limitations. Such exercises in one-upsmanship divert energies from the true issue--the costs of regulation. These costs are better illustrated by appealing to the theory and reality of competitive markets and the empirical reporting of desirable results occurring in less regulated markets.

Many other authors have estimated some subset of the lists of costs (as cited in the table) on different data bases. Friedlaender (1969) estimates the cost of rail pricing which is higher than the marginal costs of moving the goods for four aggregate commodities to be between $200 and $400 million per year.

Harbeson (1969) estimates the costs of inefficient resource use from the misallocation of traffic from rail to truck due to inefficient pricing to be between $1 and $2.8 billion in 1963. Boyer (1975) estimates this same loss at $125 million. Antonji (1976) suggests that this loss may be near zero. Levin and Peck (1976) suggest that this figure is $17 million for a limited subset of commodities--extrapolated to $68 million for a full set. Moore (1975) estimates the loss to be between $450 and $900 million which would grow to $.6 to $1.3 billion if the loss of traffic not carried is added. Keeler (1976) suggests that the pricing distortion costs approximate $500 million. Hilton (1973) also suggests low price distortion costs. Keeler (1976) suggests that the major cost of regulation ($3.5 billion) is distortion due to excess capacity.70/

To reiterate, numerical estimation of the benefits and costs of regulation is not advanced enough to advocate one result over another despite the fact that approximately ten economists have worked on the problem in the last ten years. The methodology of cost determination is fairly well developed. The compromises which must be made in that methodology because of the serious data problems which exist (limited and aggregate demand information and limited and aggregate cost of production information) make any attempt at cost/benefit analysis an exercise in futility.

While the Task Force agrees that costs exist due to regulation, we do so for a priori theoretic reasons and because we can observe savings which appear in less regulated markets. For the reasons stated above, we are reluctant to put a dollar value on the costs of current regulation or the dollar benefits to be brought about by regulatory reform. While having no doubt that positive benefits will accure to regulatory reform, we do not want to become involved in numerical gymnastics. We will stand on the observed performance of the unregulated markets and our own mutual judgment.

70/See Selected Bibliography, following this appendix, for specific citations of the aforementioned sources.

TABLE A.1
COST COMPARISONS BETWEEN MOORE AND THE ICC

	Moore Estimates (Billions of Dollars)			ICC Estimates (Billions of Dollars)	
	Low	Medium	High	Low	High
Type of Loss					
Inefficient Use of Mode					
Rail	1.7	2.0	2.4	0.0	0.0
Common Carrier Truck	1.4	1.7	1.9	(2.2)	1.0
Private Truck	.1	.2	1.0	0.0	0.0
Water	.2	.3	.3	0.0	0.0
Subtotal	3.4	4.2	5.6	(2.2)	1.0
Modal Choice	.2	1.1	2.9	(0.4)	0.0
Traffic not carried	.2	.3	.4	(0.2)	0.0
Total Estimated Loss (1968 base)	3.8	5.6	8.9	(2.8)	1.0
Total Estimated Loss (1975 base)	6.5	9.6	15.2	(4.8)	1.7

( ) denotes benefit

Selected Bibliography for Appendix A

Allen, W. Bruce, and Edward Hymson, The Interstate Commerce Commission's Staff Analysis of the Costs and Benefits of Surface Transport Regulation, Executive Office of the President, Council on Wage & Price Stability, CWPS-229 (Wash., D.C., January 19, 1977).

Altonji, Joseph, "Estimating Misallocation of Traffic Between Rail and Truck Transport," in Proceedings --17th Annual Meeting, Transportation Research Forum, Vol.XVII, No. 1, (Oxford, IN: Richard B. Cross Co. 1976), pp. 378-387.

Boyer, Kenneth, The Price Sensitivity of Shippers' Mode of Transport Selection and the Intermodal Allocation of Freight Traffic, unpub. Ph.D. dissertation, Econ. Dept. (U. of MI, Ann Arbor, MI, 1975).

Bureau of Economics, A Cost and Benefit Evaluation of Surface Transport Regulation, ICC, Statement No. 76-1 (Wash., D.C., 1976).

Capron, William, Technological Change in Regulated Industries (Washington, D.C.: Brookings, 1971).

Friedlaender, Ann, The Dilemma of Freight Transport Regulation, (Washington, D.C.: Brookings, 1969).

Friedlaender, Ann, "The Social Costs of Regulating the Railroads," American Economic Review, Vol. 61 (May, 1971) pp. 226-234.

Harbeson, Robert, "Towards Better Resource Allocation in Transport," Journal of Law and Economics, Vol. 12, No. 2, (October, 1969) pp. 321-338.

Hilton, George, "The Costs to the Economy of the Interstate Commerce Commission," in U.S. Congress, Jt. Econ. Comm., Subcom. on Priorities and Economy in Gov't., The Economics of Federal Subsidy Programs, Part 6. Trans. Studies, Submitted Feb. 26, 1973 (Washington, D.C.: U.S.G.P.O., 1973).

Keeler, Theodore, On the Economic Impact of Railroad Freight Regulation, Working Paper No. SL-7601, Econ. Dept. U. of Calif., Berkeley, CA (Sept. 7, 1976).

Levin, Richard, and Merton Peck, "Allocation in Surface Freight Transportation: Does Rate Regulation Matter?", Dept. of Economics Discussion Paper No. 31, Yale Univ., New Haven, CT (Aug., 1976).

Moore, Thomas Gale, Freight Transportation Regulation (Wash., D.C.: Amer. Enterprise Institute for Public Policy Research, 1972).

Moore, Thomas Gale, "Deregulating Surface Freight Transportation," in Almarin Phillips, ed., Promoting Competition in Regulated Markets (Wash., D.C.: Brookings, 1975) pp. 55-98.

Task Force on Railroad Productivity, Improving Railroad Productivity, Nat'l. Commission on Productivity and Council of Economic Advisors (Wash., D.C., November, 1973).

Appendix B

EMPIRICAL EXAMPLES WHERE WHAT WE PROPOSE IN REGULATORY REFORM HAS ALREADY BEEN IMPLEMENTED

A number of examples exist where the type of regulatory reform which we advocate has already been put to the market test. In most cases the results tend to support the theoretical reasoning that more competition will lower prices and increase service levels.

These examples include:

1. The intra-Texas experience of Southwest Airlines.
2. The intra-California experience of Pacific Southwest Airlines (PSA).
3. The experience of commuter airlines.
4. Barge movements under the bulk exemption.
5. Truck movements in the commercial zone.
6. Truck movements in the exempt zone surrounding an airport for incidental to air movements.
7. Exempt agricultural movements by truck.
8. Australian trucking experience.
9. Canadian trucking experience.
10. Lessons from various European countries with respect to regulatory reform.

11. Truck movements in New Jersey; and

12. Experience under the Railroad Revitalization Regulatory Reform Act.

Southwest Airlines is an intra-Texas carrier. It has been operating since 1971 and has grown rapidly competing with CAB regulated truckline carriers, e.g., Braniff, and CAB regulated local service carriers, e.g., Texas International, in markets such as Dallas-Houston, Houston-San Antonio, Dallas-San Antonio.

By virtue of its intrastate nature, Southwest is exempt from CAB regulation. The Texas Aeronautics Commission, which regulates intrastate air in Texas, has restricted entry but not fare levels. As a result Southwest offers the lowest seat mile fares in the country. Their weekday fares are lower than their CAB regulated competitors for comparable hauls in other markets by 30 to 40 percent, and their weekend and night fares are 37 to 40 percent lower than the above fares (Simat, et al., 1976, p. 6). Their strong growth rate has shown the public response to this low cost, frequent service. See Simat, Helliesen, and Eichner (1976); Harvard Business School Case Study (1975);MacAvoy and Snow, ed., (1977a); Committee on the Judiciary (1975); and American Enterprise Institute (1977) for further details.

Similar experience is found in the California market, especially prior to 1965. The State of California Public Utilities Commission basically allowed free entry and virtually flexible pricing. PSA today charges approximately half of what an interstate carrier would charge in a comparable market, e.g., 1974 market prices Los Angeles-San Francisco: $18.75 for 338 miles; Chicago-Minneapolis: $38.89 for 339 miles; New York-Pittsburgh: $37.96 for 335 miles. PSA runs at higher load factors than do the regulated airlines, gets better equipment utilization, and obtains higher labor productivity with unionized labor. For a discussion of the PSA experience, see Jordan (1970); Committee on the Judiciary (1975) MacAvoy and Snow (1977a); and Amer. Enterprise Institute (1977).

Commuter airlines operate under an exemption from the Civil Aeronautics Board. They must fly planes certified to hold 30 passengers or less, and subject to a net payload of 7,500 pounds or less. Upon notifying the CAB that one is entering a market, aside from minor data reporting, the commuter/air taxi operation is not subject to any fare or entry/ exit constraints. Approximately 130 commuter airlines (schedules with at least five round trips per week) operate in the United States today.

Some fly routes abandoned by CAB regulated carriers, e.g., Altair flies Richmond to Philadelphia once flown by National. Some fly as substitute service for regulated carriers, e.g., Atlantic City Aviation flies from Philadelphia to Atlantic City as an Allegheny commuter. Some fly in city pairs which are flown by regulated carriers, e.g., Philadelphia-New York, Philadelphia-Washington, but most fly in city pairs markets which never had or lost regulated service.

The usual experience is that commuter fares are lower than comparable regulated fares and that service is much more frequent. These results are probably due to cheaper, nonunionized labor, and smaller, more economical for the city pair, segment planes as well as higher load factors. While the quality of service is not as high as with jet operation with a regulated carrier, the impressive growth of commuter operations is testimony to the public's acceptance of this type of service. For more detail see USDOT (1976).

Virtually 90 percent of all interstate barge movements are exempt by virtue of Section 303(b) of the Interstate Commerce Act. This is the so-called commodity exemption, i.e., whenever three or fewer commodities are shipped in bulk in a single tow, there is no economic regulation of the movement. There was no enforcement of the constraint of three such commodities, and this constraint has been recently removed. No history of complaints exists because of this statute. In fact, it had to be renewed several years ago, and no opposition to renewal was expressed. The industry establishes long-term contract rates and practices rational peak-load pricing. The fact that no complaints are heard would seem to suggest that shippers/receivers are satisfied with the performance of the market. See USDOT (1973) for more information.

Under Section 203(b)8 of the Interstate Commerce Ace, provision is made to establish areas surrounding incorporated municipalities in the U.S. where motor carriers are exempt from economic regulation. By virtue of a population mileage formula established by the Interstate Commerce Commission after WWII and revised

1977, an exempt zone from three to twenty miles surrounds every incorporated municipality.

When such municipalities are adjacent to state borders, e.g., Philadelphia, Chicago, St. Louis, Cincinnati, etc., such exempt zones entail the normally regulated interstate shipments. In addition, the ICC has treated motor carrier terminal areas as coincidental with the commercial zone, e.g., a carrier certificate to serve Philadelphia can serve the whole Philadelphia commercial zone (unless specifically prohibited by its certificate). As such expansion of a zone means de facto entry into the long haul market by carriers serving the base city of the zone.

Research done on a before and after nature on areas which were previously outside of a zone and then admitted into the zone shows rate decreases from five to 30 percent after admission to the zone, reduced interlining, i.e., more single line truck service leading to a reduction in shipping time, and better service due to more competitors available to move the products. (See Allen in U.S.D.O.T., 1975). Spanton (1972) documents the situation in Baltimore, while the Allen material above describes the Philadelphia experience. The U.S. General Accounting Office is monitoring the 1977 increase in the commercial zones of all U.S. incorporated places.

Movements by truck which will have subsequent movements by air or which had previous movements by air in an approximate 25-mile radius from the airport are exempt from ICC economic regulation. The lack of complaints regarding service within the area is evidence that such service is satisfactory. The ICC in 1977 has announced hearings to investigate possible expansion to 100 miles.

Under Section 203(b)6 of the Interstate Commerce Act, movement of unprocessed agricultural products by motor carrier is exempt from ICC regulations. From time to time, the definition of unprocessed has changed and commodities have moved from being regulated to being unregulated. Winter and Ulrey of the U.S. Department of Agriculture (1961) have documented those situations and done a before-and-after study on rates. As reported in Moore (1975) and Fellmeth (1970), the rates fell on the average by 20 percent. Miklius and Casavant (1977) show that such service entails considerable stability and is not dominated by "fly by night" operations, service debilitation, and pressure for regulation in this market segment. Their conclusion was questioned by the audience at a recent conference on trucking and regulatory reform sponsored by the National Academy of Sciences and Northwestern University in April, 1977. Another viewer from an agricultural state stated that service was "just fine." The transcript and reports given at that conference will be published by the National Academy of Sciences.

Australia deregulated its trucking industry several years ago. A U.S.D.O.T. report by Canellos (1975) shows that service and rates are regarded as favorable by Australian shippers and transportation experts after deregulation, although concentration has increased in the trucking industry.

Some Canadian provinces do not regulate trucking. Sloss (1970, 1976) has investigated this situation and concluded that the unregulated trucking rates were 6.73 percent below those which were regulated, ceteris paribus (1970). Sloss resubstantiates these conclusions in the 1976 article.

Thomas Gale Moore (1976) reports on lessons of European experience with regulation in England, Belgium, Germany, the Netherlands and Sweden. In some European markets Moore points out that rates fell 40 to 50 percent after deregulation (p. 145). Service levels also improved.

Wyckoff (1974) reports that after deregulation in England truckload rates fell 10 percent while less-than-truckload (LTL) rates increased 40 percent. However, Wyckoff failed to adjust such results for a 20 percent inflation which struck England at the time. Adjusting for this factor, truckload rates fell by 30 percent in real terms and LTL rates increased by 20 percent. (For an enlarged discussion of the Wyckoff data, see Allen and Hymson, 1977.) Since more tonnage is truckload that LTL, the net result of the above is a price decrease for tonnage.

Current research by Allen (1977) concerning the intrastate trucking industry in New Jersey (which is totally unregulated with respect to rates and entry/exit) shows service equal to or better than comparable interstate movements and rate levels typically 13 to 22 percent below the Middle Atlantic tariff (the governing tariff for interstate movements from/to New Jersey). In some cases repetitive moves by the same carrier are 60 to 70 percent below what a shipper was paying a previous carrier. Some carriers publish tariffs of their own, some discount the Mid-Atlantic tariff, and some rates are negotiated "on the back of an envelope."

There has been no legislative pressure to regulate the New Jersey industry since 1961 except for household moving and dump trucks. In fact, in the last twenty years, aside from regulation of household moving and dump trucks, only two attempts to regulate the industry or segments thereof were made. Both measures were not brought to vote in the legislative bodies. Such experience suggests that the legislators perceive very little or no pressure from their constituents to regulate. Discussion with approximately 50 traffic managers of firms employing over 250 individuals did not reveal a single traffic manager who wanted increased trucking regulation in New Jersey.

While very little application of the regulatory reform provisions of the Railroad Revitalization and Regulatory Reform Act of 1976 has taken place, e.g., peak load pricing, distinct service rates, zone of reasonableness seven percent rate yo-yo, the two recent reports by the Secretary of Transportation (1977) and the Chairman of the ICC (1977) to Congress show no adverse impacts of the reforms enabled in the Act. The Secretary of Transportation, in fact, suggests that ICC procedures and interpretation have inhibited the carriers from responding to the reforms.

These instances clearly demonstrate the positive benefits of the type of regulatory reforms which we advocate in situations which are analogous to the situations with which we are dealing.

Selected Bibliography for Appendix B

Allen, W. Bruce and Edward Hymson, Comments on the Interstate Commerce Commission's Staff Analysis of the Costs and Benefits of Surface Transport Regulation, Council on Wage and Price Stability, Executive Office of the President, CWPS-229 (Washington, D.C., 1977).

American Enterprise Institute for Public Policy Research, Competition in the Airlines: What Is in the Public Interest? Roundtable Discussion (Washington, D.C. 1977).

Canellos, Georgia, Motor Carrier Regulation and the Dynamics of Interstate Freight Transport in Australia, unpub. U.S. Trans. Dept. Paper (Washington, D.C., July 1975).

Chairman of the ICC, The Impact of the Railroad Revitalization and Regulatory Reform Act to Railroad Ratemaking Provisions (Washington, D.C.: ICC, October 5, 1977).

Committee on the Judiciary of the U.S. Senate, Civil Aeronautics Board Practices and Procedures, Report of the Subcommittee on Administrative Practice and Procedure (Wash., D.C.: U.S.G.P.O., 1975).

Douglas, George and James C. Miller, Economic Regulation of Domestic Air Transport (Wash., D.C.: Brookings, 1974).

Fellmeth, Robert, The Interstate Commerce Omission (New York: Grossman, 1970).

Harvard Business School Case Study, Southwest Airlines (A), Copyright by the President and Fellows of Harvard College, Cambridge, Mass. 9-575-060, Rev. May 1975, (1975).

Jordan, William, Airline Regulation in America (Baltimore: Johns Hopkins Press, 1970).

MacAvoy, Paul and John Snow, ed., Regulation of Passenger Fares and Competition Among the Airlines (Wash., D.C.: Amer. Enterprise Institute for Public Policy Research, 1977a).

MacAvoy, Paul and John Snow, ed., Regulation of Entry and Pricing in Truck Transportation (Wash., D.C.: Amer. Enterprise Institute for Public Policy Research, 1977b).

Miklius, Walter and Kenneth Casavant, "Stability of Motor Carriers Operating Under the Agricultural Exemption," in Paul MacAvoy and John Snow, ed., Regulation of Entry and Pricing in Truck Transportation (Wash., D.C.: Amer. Enterprise Institute for Public Policy Research, 1977).

Moore, Thomas Gale, "Deregulating Surface Freight Transportation" in Almarin Phillips, ed., Promoting Competition in Regulated Markets, (Wash., D.C.: Brookings, 1975) pp. 55-98.

Moore, Thomas, Trucking Regulation: Lessons from Europe (Wash., D.C.: Amer. Enterprise Institute for Public Policy Research, 1976).

Secretary of Transportation, "Response to Congress on Section 202d of the Railroad Revitalization Regulatory Reform Act," (Wash., D.C., Oct. 1977).

Simat, Helliesen, and Eichner, An Analysis of the Intrastate Air Carrier Regulatory Forum: Vol. I, Summary Report, prepared for the Air Transportation Policy Staff, Office of the Sec'y., U.S. Dept. of Trans. (Wash., Jan. 1976).

Sloss, James, "Regulation of Motor Freight Transportation: A Quantitative Evaluation of Policy," Bell Journal of Economics and Management Science, Vol. 1 (Autumn, 1970) pp. 334-347.

Sloss, James, "Regulation of Motor Freight Transportation in Canada, A Reappraisal of Policy," Transportation Research Board Special Report No. 168 (Wash., D.C.,1976) pp. 56-65.

Spanton, Donald, Commercial Zone Expansion: Incremental Deregulation of the Motor Carrier Industry, unpub. Ph.D. Dissertation (Washington, D.C., The American University, 1972).

U.S. Dept. of Agriculture, Livestock Trucking Services: Quality, Adequacy, and Shipment Patterns, Economic Research Service, Agricultural Econ. Report No. 312 (Washington, D.C., October 1975).

U.S. Dept. of Transportation, The Barge Mixing Rule Study, TPI 30 (Wash., D.C., 1973).

U.S. Dept. of Transportation, before the ICC, "Comments of the United States Department of Transportation on Commercial Zones and Terminal Areas," Notice of Proposed Rulemaking ex Parte No. MC-37 (Sub. No. 26) (Wash., D.C., 1975).

U.S. Dept. of Transportation, Air Service to Small Communities, Comments of the U.S.D.O.T. before the CAB (Wash., D.C., July 19, 1976).

Winter, J.C. and Ivan Ulrey, Supplement to Interstate Trucking of Frozen Fruit and Vegetables under the Agricultural Exemption, U.S. Dept. of Agr. Mrktg. Research Div., Supplement to MRR-316 (July 1961).

Wyckoff, Daryl, "Motor Carrier Deregulation--Some Unanswered Questions," speech before the Jt. Mtg. of the Ohio Chapter of the Trans. Research Forum and The Eastern Central Motor Carriers Assn. (Dec. 17, 1974).

Appendix C

A NOTE ON STUDIES OF TRANSPORTATION SYSTEMS CAPACITY IN MEETING COAL TRANSPORTATION DEMAND

Several studies have been conducted which attempt to identify capacity limitations of the transportation system in meeting the growing demand for coal shipment. While the Task Force has neither evaluated nor endorsed the results of these studies, they are cited for the information of the other Task Forces within the project. As noted in the Introduction above, the Task Force did not seek to identify the physical constraints of the system. Not only is such a venture beyond the means of the Task Force, it is fraught with uncertainties which must be resolved prior to any meaningful conclusions, and which constitute the core of the Task Force effort. It is the position of the Task Force that, when the proper ground rules are in place, investments in needed facilities will be made: capital will flow to its productive uses. Defining those ground rules is thus a prior, and more important task than the prediction of system inadequacies; once institutional constraints to the expansion of the system are identified and removed, the physical constraints will become temporary.

The two studies which seek to identify these constraints explicitly recognize the considerable uncertainty of their findings. One report, performed by Input Output Computer Services for the U.S. Department of Transportation (December 1976), included several policy recommendations such as: a balancing of the regulatory treatment of various modes, either by increasing rail subsidies or by decreasing highway and waterway subsidies, granting the railroads permission to negotiate contract rates, reform of work rules, and the implementation of a more flexible pricing system.

Another study, performed by Manalytics, Inc., of San Francisco (Project 437-1, September 1976) for the Electric Power Research Institute, sought to define system capacity based on a model developed by the Federal Rail Administration. System capacity is defined as the load which results in "intolerable" delays. The report concluded that increased coal traffic would overload the rail system from 17 to 25 rail links, and that many of the waterways which carry coal currently function at or near full capacity. These results, however, are subject to critical changes in the ground rules of public policy. Waterway user charges, for example, would reduce traffic on the waterways and would also induce barge lines to arrange their tows for more expedited handling at docks, which are key choke points.

This study concludes that the fixed rights-of-ways, not barges or rolling stock, are the likely physical constraints on the system. Since investment in these fixed facilities hinges on public policy, the question of increasing system capacity may take two basic forms: where are subsidies needed to expand capacity, or what policies must be changed to make more efficient use of existing facilities and to allow investment in additional capacity where demand warrants expansion. Having chosen the latter approach, the Task Force has found these studies to be more useful as an indication of the system malaise than as a direct guide to policy.

A third study, in the form of a Report to the Secretary, is being prepared by the Coal Transportation Task Force of the Department of Transportation, and has appeared in preliminary form. This report is entitled, Transporting the Nation's Coal--A Preliminary Assessment Report to the Secretary of Transportation from the Coal Transportation Task Force, U.S. Department of Transportation (January, 1978). This study also attempts to project future coal movements and to identify system bottlenecks.

Selected Bibliography for Appendix C

Allen, W. Bruce, and Edward Hymson, The Interstate Commerce Commission's Staff Analysis of the Costs and Benefits of Surface Transport Regulation, Washington, D.C.: Executive Office of the President, Council on Wage and Price Stability, CWPS-229,(Jan. 19, 1977).

Altonji, Joseph, "Estimating Misallocation of Traffic Between Rail and Truck Transport," in Proceedings --17th Annual Meeting, Transportation Research Forum, Vol. 18, No. 1,(Oxford, IN: Richard B. Cross Co., 1976) pp. 378-387.

Bechtel Corporation, "Clean Coal Energy: Source to Use Economics, Phase I." ERDA Report, FE-1552-T1, (May, 1975).

Bhutani, J., et al., "An Analysis of Constraints on Increased Coal Production," Dept. of Interior, Contract No. 14-01-0001-1937, Mitre Report MTR 6830, January, 1975).

Boyer, Kenneth, The Price Sensitivity of Shippers' Mode of Transportation Selection and the Intermodal Allocation of Freight Traffic, unpub. Ph.D. dissertation, Econ. Dept., Univ. of MI, Ann Arbor, MI: (1975).

Bureau of Economics, A Cost and Benefit Evaluation of Surface Transport Regulation, Washington, D.C.: ICC, Statement No. 76-1, (1976).

Congressional Research Service, "National Energy Transportation, Volume I - Current Systems and Movements," Publication No. 95-15, Wash., D.C.: (May, 1977).

Friedlaender, Ann, The Dilemma of Freight Transport Regulation, (Wash., D.C.: Brookings, 1969).

Friedlaender, Ann, "The Social Costs of Regulating the Railroads," American Economic Review, Vol.61, (May, 1971), pp. 226-234.

Harbeson, Robert, "Towards Better Resource Allocation in Transport," Journal of Law and Economics, Vol. 12, No. 2, (October, 1969), pp. 321-338.

Hilton, George, "The Costs to the Economy of the Interstate Commerce Commission," in U.S. Congress, Jt. Econ. Comm., Subcom. on Priorities and Economy in Gov't., The Economics of Federal Subsidy Programs, Part 6. Transportation Studies, Submitted Feb. 26, 1973, (Wash., D.C.: U.S.G.P.O., 1973).

Input Output Computer Services, Inc., "Water Transportation Requirements for Coal Movement in the 1980's," (Wash., D.C.: U.S. Dept. of Trans., PB 263 368, December, 1976).

Keeler, Theodore, On the Economic Impact of Railroad Freight Regulation, Working Paper No. SL-7601, Dept. of Economics, Univ. of Calif., Berkeley, (Sept. 7, 1976).

Levin, Richard, and Merton Peck, "Allocation in Surface Freight Transporation: Does Rate Regulation Matter?", Dept. of Economics Discussion Paper No. 31, Yale Univ., New Haven, CT, (August, 1976).

Moore, Thomas Gale, "Deregulating Surface Freight Transportation," in Almarin Phillips, ed., Promoting Competition in Regulated Markets, (Wash., DC.: Brookings, 1975), pp. 55-98.

Moore, Thomas Gale, Freight Transportation Regulation, (Wash., D.C.: American Enterprise Institute for Public Policy Research, 1972).

Task Force on Railroad Productivity, Improving Railroad Productivity, (Wash., D.C.: Nat'l Commission on Productivity and Council of Economic Advisors, November, 1973).

Appendix D

SITUATION REVIEW: TRANSPORTATION OF ENERGY EQUIVALENT IN COAL BY HIGH VOLTAGE TRANSMISSION LINES

I. INTRODUCTION*

All aspects of mine mouth generation and water supply and air pollution problems created by power plants are not handled in this report. This paper specifically addresses the transmission of electricity after it has been generated as one aspect which has not yet been discussed.

At least one engineering firm is of the opinion that "of the four alternative modes of transporting energy (transmission lines, railroads, railroads plus barges, slurry pipelines) transmission lines probably have the greatest environmental impact."(3) Unlike the other alternatives, transmission lines require a substantially large power plant to account for I^2R losses.

Across the U.S. it has been estimated that the transmission of electricity costs amount to about 13 percent of the delivered cost and distribution costs amount to about 37 percent.(6) Within this basic framework, major variables exist through which the efficiency of energy movement can be increased. It has been demonstrated that high voltages transmit more electricity over longer distances at less cost. A single 765-kV line can carry as much power as thirty 138-kV lines. The 765-kV line required only

*The numbers in parenthese in this Appendix refer to the documents listed on the page entitled "References" following this Appendix.

only 1/13th the land area per kilowatt of capacity. Furthermore, construction costs per kilowatt of higher-voltage line are typically 10 times less than those of lower-voltage lines. The benefits of higher voltages include conservation of land, fuel, and funds, and more efficient and reliable service at lower rates.[9]

We are presently building overhead transmission lines at the rate of 10,000 miles per year. These lines require approximately 20 acres per mile. There are more than 200,000 miles of lines rated 138-kV or higher. The Federal Power Commission estimates that the current 14,000 circuit miles of 500-kV and higher lines will increase to 24,000 miles by 1980 and to 42,000 miles by 1990. Experimental AC and DC lines in the 1200 kV range are also being tested. AC lines are considered to be proven at 765-kV with a reasonable chance of success at 1100-kV.[3] However, AC lines are subject to instabilities at great distances.[1] For DC lines, conversion costs are high and the technology may be limited to less than 600-kV.[1,3]

II. ENVIRONMENTAL ISSUES

A. Electric Field Effects

Electric fields have only recently been identified as an area of concern. One possible reason for this concern is the occurrence of smaller insulation clearances paralleling increases in line voltage, as is reflected in lower towers and line-to-ground clearances. The principle concerns in this area include shock and biological effects.

1. Shock

In working near a high-voltage line, the shock is comparable to that which occurs as you walk across a rug and touch a doorknob. The electrical discharge is small, however, the involuntary reaction is startling and can be hazardous, depending on the circumstances.

2. Biological Effects

The second major concern is that electric fields may induce biological effects. Alleged illness of workers in high-voltage substations in the U.S.S.R. stirred worldwide concern. However, subsequent examinations of similar workers in Canada, England, France, West Germany, Italy, or the U.S. fail to confirm the U.S.S.R. reports.

This concern has been stimulated by hearings before the New York State Public Service Commission on the proposed construction of a 765-kV line. The possibility of biological effects has likewise been a key issue in hearings on a proposed 430-mile, ±400 DC line that would traverse portions of Minnesota and North Dakota. If electric fields cause biological effects, it appears that they are subtle and extremely difficult to identify. Since we know that higher voltage enables a dramatic increase in right-of-way utilization, agreement on field effect standards is very important. Additional research is undoubtedly needed. EPRI's extensive research program, together with ERDA's efforts in this regard should resolve this issue.

B. Audible Noise

The second environmental consideration associated with transmission is the problem of noise control. Corona phenomena associated with high voltages have been the subject of research for many years. By selection of conductor size and phase spacing on the line, corona generation can be controlled. This and other engineering aspects of noise are understood, including its correlation with rain and fog. What is not understood are its psychological effects: at what level do people become annoyed with the noise from a transmission line? The answer has direct implications for effective right-of-way use. Audible noise at the edge of a right-of-way may govern the final design and economics.

C. Aesthetic Design

The third environmental consideratin of overhead transmission is line design. This issue, relating to aesthetics, while not totally equivalent to the underground transmission issue, is contained within the question and is, therefore, part of the driving force towards consideration of that method. This issue is discussed in the following section.

III. UNDERGROUND TRANSMISSION

Underground transmission has existed since the days of Edison, but only one percent of all lines are now buried, with 19th century design still being used in most cases.(8) Cost is the paramount reason for such scant use of underground lines. Today 12 to 16 times more initial capital investment is needed, and repairs on underground cables are expensive and time consuming. Additionally, enclosed lines are less efficient than bare overhead conductors at most voltages. It is estimated that if all future transmission lines were to be placed underground, the cost of electricity would rise 30 to 50 percent, from $40 billion to $55-60 billion. If existing lines were also transferred underground, electricity bills would be at least doubled.

Nevertheless, transmission in the future may have to move away from overhead lines, as they are aesthetically objectionable and they require that rights-of-way be sought. These rights-of-way are becoming increasingly costly in congested areas. The sensitivity of the public to environmental issues has brought about a demand for more attractive T&D lines and more productive use of rights-of-way.

One writer states that a strong case can be made for a positive commitment to undergrounding certain power facilities, while others could remain aboveground.(7) For example, lines and substations serving industrial demands, which account for 40 percent of the utilities' load would not be laid underground.

EPRI, in cooperation with ERDA, has begun the largest R&D program ever undertaken ($40 MM this year) on transmission and distribution. This R&D is balanced between overhead and underground programs. A further review of selected R&D efforts indicates a wide range of opportunities exist for improvement of the underground delivery of power.

REFERENCES FOR APPENDIX D

(1) National Academy of Engineers Report, "U.S. Energy Prospect," (1974).

(2) Bridges, J.E., "Biological Effects of High Voltage Electric Fields: State-of-the-Art Review and Program Plan," ITT Research Institute Report ITTRI Project E8151, (November 1975).

(3) EBASCO, "Energy Transportation," (April 1976).

(4) Upper Midwest Council, "Northern Great Plains Coal," (April 1976).

(5) Young, F., "Packing the Power Corridor," EPRI Journal, (May 1976).

(6) U.S., GAO, Report RED-76-107, "Status of Federal and Private R&D Efforts to Conserve Energy by Reducing Electric Power Transmission Losses," (June 1, 1976).

(7) Perry, E.R., "T&D: The Measured Move Underground," EPRI Journal, (September 1976) p. 6.

(8) Public Utilities Fortnightly, "Era of Underground Transmission is Approaching," (April 14, 1977) p. 63.

(9) Kornberg, H., "Concern Overhead," EPRI Journal, (June 1977).

Appendix E

HIGHWAY MAINTENANCE ATTRIBUTABLE TO TRUCKS

The Motor Vehicle Manufacturers Association (MVMA) publication Motor Vehicle Facts and Figures 76, lists 1974 passenger car miles between cities on page 54.

Inter-city passenger car travel = 1.143×10^9 miles.

The same publication lists ton-miles for inter-city truck freight on page 56.

Truck travel = 441×10^9 ton-miles in 1974.

The American Trucking Association estimates tonnage hauled by trucks is

784×10^6	tons by regulated trucks
$1,131 \times 10^6$	tons by non-regulated trucks
$1,915 \times 10^6$	tons total truck tonnage.

Average truck hauling distance is

$$\frac{441 \times 10^9}{1,915 \times 10^9} = 230 \text{ miles.}$$

Assume the average truck load is 25 tons:

$$\frac{441 \times 10^9}{25} = 17.6 \times 10^9 \text{ truck miles in 1974.}$$

Studies indicate that highway damage increases as the cube of the weight ratio, so that a 55,000-pound truck will be equivalent to 2,500 passenger cars:

$$\frac{(55,000)^3}{(\ 4,050)} = 2,500.$$

Therefore, equivalent passenger car miles for truck mileage is

17.6×10^9 truck miles x 2,500	=	44.0×10^{12}
Actual passenger car miles	=	1.1×10^{12}
Total		45.1×10^{12}

Percent of total equivalent miles contributed by trucks:

$$\frac{44.0 \times 10^{12}}{45.05 \times 10^{12}} \times 100 = 97.6 \text{ percent.}$$

The Federal Highway Administration estimates the total highway maintenance budget is $\$7.0 \times 10^9$ per year.

Proportion of highway maintenance budget attributable to trucks:

$$.976 \times 7.0 \times 10^9 + \$6.84 \times 10^9 \text{ per year.}$$

Coal hauling is 66 million tons with an average haul of 60 miles:

$$\frac{66 \times 10^6 \times 60}{25} = .1584 \times 10^9 \text{ truck miles.}$$

Percent of total hauling attributable to coal trucks:

$$\frac{.1584 \times 10^9}{17.6 \times 10^9} \times 100 = .9 \text{ percent.}$$

Highway maintenance required for coal trucks:

$$\$6.84 \times 10^9 \times .009 = 62 \times 10^6$$

Ton-miles for coal hauling are

$$66 \times 10^6 \times 60 = 3.96 \times 10^9$$

Cost per ton mile is

$$\frac{62 \times 10^6}{3.96 \times 10^9} = \$.016/\text{ton-mile.}$$

The added maintenance cost for the average 60-mile haul is

$$\$.016 \times 60 = \$.96/\text{ton of coal.}$$

Added road maintenance incurred by increased coal production in 1990:

$1,300 \times 10^6 \times .08 = 104 \times 10^6$ tons transported by truck;

$104 \times 10^6 \times \$.96 = \$100 \times 10^6$

Truck taxes for 55,000 GVW trucks are $2,691 annually. (From page 91 of MVMA Facts and Figures 76.)

A truck can make three 60-mile round trips per day, so one truck can make

$$3 \times 250 = 750 \text{ trips/year.}$$

Number of trucks required to haul 104 million tons of coal is

$$\frac{104 \times 10^6}{25 \times 750} = 5,550 \text{ trucks.}$$

Taxes paid by trucks

$$5{,}550 \times \$2{,}691 = \$14.9 \times 10^6$$

Net maintenance not covered by taxes is

$$100 \times 10^6 - 14.9 \times 10^6 = \$85.1 \times 10^6$$

Cost per ton adjusted for taxes is

$$\$.96 \times \frac{85.1}{100} = \$.82 \text{ per ton of coal}$$

or

$$\$.016 \times \frac{85.1}{100} = \$.014/\text{ton-mile.}$$

List of Acronyms and Abbreviations

TRANSPORTATION TASK FORCE

CAB -	Civil Aeronautics Board
DOT -	Department of Transportation
EDF -	Environmental Defense Fund
EPA -	Environmental Protection Agency
FWPCA -	Federal Water Pollution Control Act
ICC -	Interstate Commerce Commission
LTL -	Less Than Truckload
NCPP -	National Coal Policy Project
NEPA -	National Environmental Policy Act
O & M -	Operation and Maintenance
STA -	Slurry Transportation Association
WPA -	Water Protection Agency
4-R Act -	Railroad Revitalization and Regulatory Reform Act, 1976

Glossary

TRANSPORTATION TASK FORCE

andromous fish losses - loss of fish during upriver spawning runs.

aquifer - water-bearing stratum of permeable rock, sand, or gravel.

aquifer compaction - as aquifers are overdrawn, ground water storage capacity can be reduced, producing irreversible damage.

bulk commodity exemptions - Part 3, Section 303(b) of the Interstate Commerce Act exempts from rate regulation barge movements of three or fewer bulk goods in a single tow.

channelization - the process of channeling a stream such that drainage occurs more rapidly.

coal slurry pipelines - pipelines which transport coal mixed with water or some other liquid to provide a transportation medium.

common carriers - a legal concept which defines the duties of certain types of transportation modes. According to Part 1, Sections 104 and 106, of the Interstate Commerce Act, they are to provide transportation upon request at reasonable rates and to establish through routes with other common carriers. They also must publish tariffs.

eminent domain - right of condemnation; usually supplied by statute.

internalization of costs - the process whereby the cost/benefit is transferred from those not engaged in the economic activity to those so engaged.

market dominance principle - railroad rates must be just and reasonable according to the Interstate Commerce Act, Part 1, Section 105(b). No railroad rate shall be found unjust or unreasonable if the railroad does not exert market dominance over a particular service. In the language of the statute, market dominance ". . .refers to an absence of effective competition from other common carriers or modes of transportation for the traffic or movement to which a rate applies."

overdraft - the amount of water extracted which exceeds the replacement from both natural and imported sources.

point source - an identifiable source of pollution as opposed to a non-point source. For example, a smoke stack is an identifiable source, whereas runoff from fields is not.

peak-load pricing - a tariff design whereby the price to consumers is made higher during time intervals in which utilities experience highest demand than during other periods of time.

riverine ecosystems - the functioning together of the total biologic community associated with the river habitat and the nonliving environment.

turbidity - opaqueness caused by stirred-up sediment, reducing transparency for light penetration in water.

unit train - a train consisting of one type of cargo designated for one specific location (usually coal).

variable costs - used to designate those costs which fluctuate with the volume or level of production.

wetlands - land or areas having continually high levels of soil moisture such as tidal flats or swamps.

SECTION 2
AIR POLLUTION TASK FORCE REPORT

TASK FORCE MEMBERS:

J. James Roosen,
Co-Chairman

Bruce J. Terris
Co-Chairman

David M. Anderson
John R. Bartlit

Joseph J. Brecher
Jackson B. Browning
Thomas L. Montgomery
Jerrel D. Smith
Michael D. Williams

Alternates:
Richard Porter

TECHNICAL STAFF:

Nathalie V. Black
Suellen T. Keiner

TABLE OF CONTENTS

AIR POLLUTION TASK FORCE

I. Introduction

The purpose of the National Coal Policy Project is "to permit the increased utilization of coal in an environmentally acceptable manner." An integral aspect of this purpose is to address the impact of coal burning on the environment. Environmental legislation and subsequent court decisions in interpreting environmental statutes and regulations over the past ten years has led to the increased use of cleaner burning liquid fossil fuels. On the other hand, the embargo on foreign oil supplies, resulting energy legislation, and increases in the cost of these energy supplies rekindled national interest in coal as a major energy source. The same issues that led to the decisions to use liquid fuel still remained. But the air pollution problems of the increased use of coal had not been solved and, in general, were perceived to be a dispute between industrialists, who wanted to keep costs to a minimum, and environmentalists, who wanted to clean up and enhance the environment. These opposing views appeared to be deadlocked with no room for reconciliation.

The Project recognized this apparent dilemma and assigned a number of these issues to the Air Pollution Task Force for examination, study, and debate. The objectives of the Task Force were to identify any possible areas of agreement, to develop compromises that would allow for increased coal utilization, and to formulate policy recommendations that could be implemented on a national scale to accomplish the purpose of the Project.

The following issues were selected as representing the most significant problems in terms of increased utilization of coal:

A. What impact will the siting of generating stations at the fuel resource location, versus the siting of generation facilities near the area of consumption, have on coal utilization, and should a public policy on such issues be established.

B. Should states adopt utility siting statutes and, if so, what provisions should be incorporated in such statutes.

C. What policy should be adopted regarding the application of Best Available Control Technology versus emission standards.

D. With regard to presently regulated pollutants, such as SO_2 and particulates, and unregulated pollutants, such as sulfates and nitrates, does the present data base (criteria) adequately support standards; what information should be available for new standards and should such new standards be set.

E. What policy can be established to recognize the unusual impact of the retrofitting of air pollution control equipment on plants with relatively short remaining life to meet emission standards.

F. Can standardized diffusion modeling be established in order to eliminate controversy and to provide a relatively consistent data base from which judgments as to the impact of the use of coal resources can be made. If so, how can such models be validated.

G. What action should be taken regarding carbon dioxide discharges resulting from the combustion of coal.

H. Can emission standards that are currently based on heat input be revised to encourage more efficient use of coal.

I. What is the proper policy regarding the siting of new plants in non-attainment areas.

J. How can air pollution policies be modified to recognize the interrelationship of air pollution, water pollution, and land use more adequately.

K. Can a pollutant tax on SO_2 be utilized to achieve compliance with ambient air standards while not interfering with increased utilization of coal resources.

L. What type of regulatory or legislative scheme should be developed with regard to the significant deterioration of air quality that will allow for reasonable and necessary industrial growth in the utilization of coal.

M. What policy can be established in order to ease the burden of non-performance of new technologies carried by industry.

One of our Task Force's major accomplishments was the establishment of a very good rapport between industry and environmental representatives, who are concerned about control of air pollution from differing perspectives. Initially it was very difficult for industry and environmental representatives to exchange ideas because the sessions operated in a formal manner and a level of trust had not been established between the two groups. However, the members quickly began to listen to and evaluate each other's viewpoints and feelings of mutual respect were established. As a result, the various issues were discussed openly and the two groups were able to agree on compromise solutions that appear to be workable if adopted as national policy.

The Task Force made significant progress towards compromise positions on the 13 issues outlined above. Compromise positions for part or all of eight issues were agreed upon. In some cases, the level of agreement is restricted to a portion of the original issue or to a more limited definition of the issue. While the Task Force was not able to agree on the remaining issues in the time available, it is possible that agreement could have been reached with additional time. The areas of disagreement on the remaining issues are discussed in a later section of this report where both groups have summarized their respective positions.

II. Summary

The issues on which agreement was reached are discussed below:

Utility Siting Statute: The Task Force agreed that states should adopt utility siting statutes which allow all siting requirements to be fulfilled through one process and hearing. We also agreed upon general provisions for a model statute. If adopted, this procedure should save considerable siting time, provide advance notice of utility plans, enable citizen groups to participate effectively, and allow for more efficient planning of new facilities.

Power Plant Siting: The Task Force agreed that power plants should be sited, to the extent practicable, near the use centers. Adoption of this policy should ultimately reduce siting requirements and improve pollution control.

New Technology: The Task Force agreed that industry should be encouraged to try new pollution control technologies and that the present legislation does not go far enough in this regard. Therefore, the Task Force agreed to a proposal which should effectively encourage industry to use new technology which will reduce emissions at lower cost.

Air Quality Standards: The Task Force agreed that a review of existing air quality criteria documents is long overdue and data on new pollutants should be developed as soon as possible. If the evidence warrants such action, new air quality standards should be adopted by the Environmental Protection Agency (EPA).

Old Power Plants: The Task Force agreed that the amount of money spent on the control of air pollution emitted from old power plants, 20 years old and older, should be the same as for the control of emissions from new plants. If such controls are not sufficient to meet the national standards, tall stacks and intermittent control systems would be permitted to achieve compliance.

Air Quality Modeling: The Task Force agreed that standardized air quality models should be used consistently. If EPA selects the best possible models, as indicated by the Task Force, this approach should greatly benefit air quality control efforts in this country.

Emission Standards Based on Output: At the present time, emission standards for electric or steam production are based on heat input. The Task Force agreed that such standards should be based on plant output such as Mwh of electricity or its equivalent. This change should promote power plant efficiency.

In addition, the Task Force members came to the following agreement. Some studies indicate that greater utilization of coal will result in significant increases in pollution emissions. If such increases seem likely, steps should be taken in air pollution control programs to minimize them, particularly in the case of sulphur oxide, particulate, and nitrogen oxide emissions. If present laws are unable to accomplish this, further amendments to the Clean Air Act may be necessary.

Carbon Dioxide: The Task Force agreed that worldwide carbon dioxide (CO_2) concentrations should be studied. Since there is much concern and little available information about this issue, a CO_2 evaluation undertaken at this time could minimize confusion and later problems.

Penalties: It was determined that this issue cut across the lines of several task forces. Therefore, a separate *ad hoc* task force on emission charges was formed with representatives from several task forces. Their recommendations are included in a separate section of the main report.

III. Policy Recommendations

The following section of this report sets forth the agreement reached after discussion and debate of the original twelve issues. During the deliberations of the task force, the format of a concise policy-oriented statement followed by a rationale was agreed upon. Each policy statement is set forth as a conclusion based upon the rationale which follows.

A. Utility Siting Statute

1. Policy: Utility siting statutes should be adopted by state governments. These statutes should require long-range planning for the construction of power units, provide for a single application procedure for obtaining all necessary state permits and approvals, require consideration of all relevant energy, economic,

environmental, and other factors, and provide for full citizen participation in all aspects of the proceedings, including grants to citizen groups (such as environmental, business, labor, or consumer organizations).

2. Rationale: Utility siting statutes would benefit industry, environmentalists, and the public generally. Such statutes would benefit industry because they would enable companies to obtain approval of long-range plans for power units, to know the siting criteria well in advance, and to obtain all necessary state permits and approvals through a single procedure during a specifically limited period of time. This should allow industry to avoid large expenditures for sites, engineering designs, and environmental studies long before there is an indication as to whether a particular site has any chance of being approved. It should also minimize situations where the states may deny approval after lengthy processing resulting in the impossibility of finding alternative sites in time to meet public requirements for electricity. It should significantly reduce the present delays in state approvals, as well as lessen public antagonism to utility siting decisions which are generally made prior to public participation or knowledge.

From the environmentalists' point of view, siting statutes are helpful because they provide notice of utility plans long before decisions are made. They allow citizens to participate in the process of utility siting from the earliest stages. They also provide for more rational environmental decision-making since long-range planning is required (all aspects of the siting decision would be considered together rather than piecemeal).

The public would be served by the offering of an opportunity to participate at an early stage in siting decisions, by general improvements in decision-making, by a reduction in both the waste of having numerous state agencies consider different aspects of the decision and the costs of such a decision to utilities which are in turn passed on to the consuming public, and by the avoidance of situations in which delays in decision-making or rejections of a project at a late stage may jeopardize the availability of an adequate supply of electricity.

The following is a description of a workable state siting statute:

The statute would require that each utility regularly submit to the state its estimates concerning future needs for electricity and its corresponding plans to meet this need. These plans should include, as early as possible, the preliminary identification of potential plant sites. The estimates and plans would be reviewed by each agency and would be subject to public scrutiny.

When the utility proposes to construct a particular power plant or energy conversion facility, except those falling below a specified size, the statute would require a certificate of public need and proof of environmental compatibility prior to construction or operation. Transmission lines, pipelines, coal gasification facilities, geothermal plants, and any associated facilities (dams, aqueducts, ponds, reservoirs, and substations) would also be covered by this requirement.

In order to obtain the necessary certificate, a utility would file a single application and would pay a fee based on the estimated cost of the facility in order to pay for the cost of processing. Public notice of the application would be issued when the application is filed. There would be provision for a single public hearing allowing comments by all related or affected state agencies, nearby governmental bodies, and any residents of the area or other affected citizens groups.

The criteria for approval or disapproval of the permit would be set forth and might include the following: energy needs, environmental impacts (including air, water, and land use), alternative site uses and sources of energy, new technology to minimize environmental impacts, monitoring requirements, and operating procedures. All state agencies having responsibility for some aspect of the decision (such as the air, water, and public service agencies) would be required to participate and make their determinations of the acceptability of the plan within their own areas of responsibility as part of the same hearing process. All federal agencies which would have responsibility concerning a proposed facility, such as EPA and the Department of Interior would be encouraged to submit their positions in a timely manner as part of these same proceedings. Time limits would be established within which all agencies would be required to make their decisions; a time limit of one year after the submission of a completed application appears to be reasonable.

Judicial review would generally be permitted only by participants in a particular siting procedure. Any other parties seeking review should be required to provide a reasonable justification for not having participated in the initial review. The time period for seeking judicial review should be strictly limited.

Citizens groups should be able to receive reasonable funding to obtain expert witnesses and attorneys so that they can participate knowledgeably and effectively in these proceedings. These groups (such as environmental, business, labor or consumer organizations) should be eligible for funding if they have (1) a position that is not already adequately represented in the proceedings and (2) either do not have funds to pay for adequate representation or are composed of members who individually do not have sufficient economic interest in the matter to justify substantial expenditures. The state should ensure that only qualified, responsible groups receive funding based on applications which assure that the funds will be properly spent. Citizen groups should be required to pay a portion of the costs of representation.

B. Criteria for Plant Siting

1. Policy: Based on air quality considerations, new coal-fired power units should be sited, to the extent practicable, in the general area where the energy is to be distributed. The application of such a policy should be consistent with generally accepted economic and engineering practices. Selection of sites not in the general area in which the energy is to be distributed shall require substantial justification. This policy of encouraging siting in areas of distribution should not apply in non-attainment areas, although units should be free, at their option and consistent with the law, to locate in such areas.

2. Rationale: Siting of coal-fired power plants can cause significant environmental impacts through pollutant emissions regardless of where they are located. These air quality impacts include the local effects of deposition of pollutants in the immediate area of the plant, and regional effects resulting from long-range transportation of pollutants, transformation of pollutants in the atmosphere, and pollutant loading due to an accumulation of emissions from the plant combining with emissions from other sources. Effects will vary depending on characteristics of the region such as whether it has a substantial population, is principally agricultural with crops sensitive to air pollution, or has especially important public land areas such as national parks.

However, in addition to environmental impacts, siting of coal-burning power plants involves social, political, economic, and regulatory factors. For example, site selection should recognize the needs to conserve petroleum products used in fuel transportation, to preserve water resources, and to protect local economies. Site selection must also be consistent with sound economic and engineering practices.

Thus, the siting of each coal-burning power unit involves a balancing of costs and benefits. In some instances, where substantial justification is shown, siting is appropriate in remote areas.

Nevertheless, as a general principle, coal-burning power units should be located in the area where the energy is to be distributed. This principle is based upon the belief that those who benefit most from the construction of new units should accept the associated social and environmental impacts. It is recognized that, in some instances, much of the energy for new units is used to produce goods and services used by those outside of the region of primary use; however, this policy best approximates imposing the social and environmental burdens on those who most benefit.

This policy is an improvement over siting in remote areas in that it encourages technological innovation in meeting more stringent controls in order to protect those larger populations living in the areas where the energy is distributed. It may also encourage energy conservation which could lead to a reduction in the number of new units required.

It is recognized that in the short run, this policy could result in the exposure of a greater number of people to higher levels of air pollution than would a policy calling for remote siting. However, this policy is justified because it is more likely to lead to better pollution control and less pollution in the future.

C. New Technology

1. Policy: EPA should be granted the authority to allow exceptions from the New Source Performance Standards (NSPS) and Best Available Control Technology (BACT) requirements of the Clean Air Act to encourage the development and use of new technology. This will require amendment of the Act so that EPA can approve a limited number of variances from NSPS and BACT (but not from compliance with the national ambient standards) in situations where there is a reasonable likelihood that NSPS will be achieved by use of the new technology. If the equipment did not immediately achieve NSPS and BACT, companies would be required to take all reasonable steps to attain those required levels of pollution control. Charges would be imposed, if the plant concerned failed to comply with NSPS and BACT, based on how closely the plant comes to meeting these standards. This requirement would be satisfied if the company made greater expenditures to reduce emissions or, alternatively, paid the charge; the former method of compliance is preferable. In no instance, however, should the charge be greater than the difference between the cost of achieving NSPS and BACT with proven technology and the lower, actual cost of the new technology used.

2. Rationale: Both industry and environmentalists have a vital interest in improving technology for pollution control. Industry is interested in meeting applicable legal requirements at lower cost. Environmentalists are interested in improving pollution controls and in lowering costs so that the country has more resources with which to improve all the various means of environmental protection.

Although the Clean Air Act Amendments of 1977 have a provision which allows EPA to grant temporary variances from NSPS for new technology, this provision is not adequate to encourage industry to utilize new technology. If the new technology fails to achieve NSPS, the provision requires the company to retrofit. As a result, the company is likely to have to pay more because it tried new technology than if it had simply used proven technology to meet NSPS. Companies are therefore unlikely to adopt new technology unless the cost savings to them are extremely large and the achievement of NSPS is almost certain. There is no incentive to try technology which will, if successful, reduce pollution substantially below NSPS where cost savings to the company are small or where the technology is unproven.

The Task Force urges that EPA be given authority to grant a limited number of permanent variances from NSPS and BACT for new technology. Such variances should be granted where a reasonable showing has been made that this unit or subsequent units will either (1) meet NSPS and BACT at a significantly lower cost than through the use of present technology, or (2) achieve significantly better pollution control than NSPS and BACT. No variances could be given from compliance with the ambient air quality standards.

Whenever EPA grants a variance, it should establish a pollution control floor for the plant based on the particular circumstances involved. If the unit does not achieve this level of pollution control, it should pay a charge equal to the difference between the cost of achieving NSPS and BACT with proven technology and the lower, actual cost of the new technology used. If the unit achieves a level of pollution control which is greater than the floor set by EPA, but less than the requirements of NSPS and BACT, a charge should be set which will be a portion of the difference between the cost of achieving NSPS and BACT with proven technology.1/ If the unit achieves the level

---

1/ We suggest the following formula: the company would pay a charge equal to the ratio of the differ-

of pollution control of NSPS and BACT, there should be no charge. Similar charges should be assessed each year based on the amount of money saved by the company in operating costs as compared to those incurred by using existing technology, and on the degree to which the system fails to achieve NSPS and BACT. If the company subsequently increases the level of removal, the charge would be partially returned, based on the proportion of the life of the unit which remained.

The company can satisfy the charge requirement (1) preferably by spending more money on improving the technology or otherwise reducing pollution emissions, or (2) if this is not done, by paying the charge. In no instance, however, should the charge, plus the cost of the new technology, be greater than the cost of achieving NSPS and BACT with proven technology.

This system of charges is intended to encourage industry to seek variances only when the new technology has a substantial

---

1/ ence between the actual emissions and emissions that would exist if NSPS and BACT were complied with, and the difference between the emissions at the EPA floor and the standard emissions if NSPS and BACT were complied with, multiplied by the cost difference between the process and the innovative system.

Symbolically the relations is:

$$\text{charge} = (Cp - Cn)\ \frac{E - Std}{E_f - Std}$$

where "Cp" represents the cost of the proven system, "Cn" the cost of the innovative system, "E" is the emissions of the innovative system, "Std" represents the emissions at the standard, and "E_f" represents the floor emissions. The following is an illustration of the relation.

Let us assume that: the unit without controls will emit 100 tons of pollutant a day; NSPS and BACT would remove 90 tons of pollutant a day so 10 tons would still be emitted; EPA, in granting the variance, determines that the appropriate floor is the removal of 70 tons of pollutant, leaving 30 tons of emissions;

likelihood of meeting the projected pollution control. It is also intended to give industry an incentive to seek to modify and improve the controls when they are not immediately successful and, if these efforts fail, to adopt other means to bring pollution control as close to NSPS and BACT as possible.

1/ the cost of achieving NSPS and BACT by proven technology is $10 million; and the actual cost of the new technology is $6 million; and the plant, after use of the new technology, removes 75 tons of pollutants and therefore emits 25 tons of pollutants.

Step 1: Subtract the level of emissions if NSPS and BACT were complied with (10 tons) from the actual emission of the new plant (25 tons), which equals 15 tons.

Step 2: Subtract the level of emissions if NSPS and BACT were complied with (10 tons) from the level of emissions at the floor set by EPA (30 tons), which equals 20 tons.

Step 3: Divide the result of Step 1 (15 tons) by the result of Step 2 (20 tons), which equals 3/4 ton.

Step 4: Subtract the actual cost of the new technology ($6 million) from the cost of proven technology to comply with NSPS and BACT ($10 million) which equals $4 million.

Step 5: Multiply the result of Step 3 (3/4 ton) times the result of Step 5 ($4 million), which equals $3 million.

The company would have to pay a charge of $3 million. However, even though it was emitting 25 tons of pollution, instead of only 10 tons if NSPS and BACT were complied with, it would still have $1 million less in costs than if it had used proven technology.

D. Air Quality Standards

1. Policy: EPA is encouraged to proceed immediately with the long overdue reexamination of the criteria documents and National Ambient Air Quality Standards (NAAQS). The Clean Air Act Amendments of 1977 call for the review and revision as necessary, of the criteria documents and NAAQS by December 31, 1980, and additional review and revision as necessary at 5-year intervals thereafter. Because of the need for review and revision of the criteria documents and NAAQS, especially for sulfur dioxide (SO_2) and total suspended particulates (TSP), this task should have a high priority and should be scheduled for completion before mid-1979 when State Implementation Plan (SIP) revisions are due. EPA should ensure that reexamination of both criteria documents and NAAQS is carried out every five years. In reviewing and revising the standard for TSP, consideration should be given to replacing the existing standard with one which takes into account the size and chemical composition of the particulates. EPA should develop adequate data on air pollution effects for new non-criteria pollutants such as nitrates, sulfates, arsenic, molybdenum, and polycyclic organic matter as soon as possible. Whenever substantial data establishes significant risk of harm to health or welfare, EPA should promptly adopt new ambient air quality standards or hazardous air pollution regulations.

2. Rationale: An example of the great need for review of the criteria documents is the TSP document, which has not been revised since it was issued in January, 1969. Many studies which have been conducted since that time cast serious doubt on the accuracy of the TSP document and conclude that it does not "...accurately reflect the latest scientific knowledge useful in indicating the kind and extent of all identifiable effects on public health or welfare which may be expected from the presence of such pollutants in the ambient air in varying quantities," as required by Section 108 of the Clean Air Act.

In reviewing the TSP standard, consideration should be given to having that standard take into account the size, distribution, and chemical composition of the particulates because most of the health impacts of TSP are caused by the smaller so-called "respirable fraction" of the dust. The larger material has a minor effect on health but a very major effect on the ambient concentration. Similarly, visibility is primarily degraded by the smaller particles. The chemical composition of TSP can vary greatly from one area to another. This variation can greatly change the effects of each material on health.

E. Old Power Units

1. Policy: Old power units -- those 20 years old and older -- should not be required to spend more money for pollution control per megawatt hour of electricity produced than is required for a new power unit. Consequently, if the expenditure of money on emission controls is not adequate to meet the ambient air quality standards as required by a State Implementation Plan (SIP), the unit should be able to use tall stacks or supplementary controls. The cost of such tall stacks or supplementary controls should be considered in determining whether an equivalent amount of money is spent as compared to that spent by new units. However, regardless of the amount of money which must be expended, the ambient air quality standards must be met. In addition, the old unit will be required to restrict total atmospheric loading (calculated as an annual average) to an amount equivalent to a new unit operating at 80 percent capacity and meeting New Source Performance Standards (NSPS). This requirement will be met by restricting the capacity factor of the old unit (calculated as an annual average), except that the unit will not be required to reduce capacity factors to an amount less than 35 percent. This policy can be selected at the option

of the old plant operator; otherwise the regular requirements of the SIP would apply.

2. Rationale and Implementation Procedure: Old units have much of their useful life behind them and are the most expensive units to control. Even without controls, this class of units would usually contribute less and less air pollution each year because of decreasing capacity factors. These units should not be required to spend more for control than new units. Moreover, when total atmospheric loading is considered, those old units are not as important as recently constructed or new units.

The recommended policy would generally not require old units to spend more for pollution control than new units. If the use of emission controls will not be adequate to meet the ambient standards, the unit could use tall stacks or supplementary controls. Total atmospheric loading would be further restricted to that of new units by reduced capacity factors.

In order to put this policy into perspective, the following information about the United States electric utility industry is useful to consider. There are 190 conventional steam units in the United States that are 20 years or older. These units comprise 566 generating units and have a total capacity of 46,000. These plants are approximately 13 percent of total 1975 conventional steam capacity in the United States. Special treatment for these units should not require a level of control more restrictive than necessary to assure that they do not significantly jeopardize the continued attainment and maintenance of ambient air quality standards. This procedure could significantly conserve capital resources and increase the utilization of the nation's coal resources. Lack of relief for many of these plants will continue to force industry toward making decisions to: (1) convert to liquid fossil fuels, (2) retire units earlier than called for and replace peaking generation with liquid fossil fuel sources, or (3) invest capital in control facilities that could be better expended and reap more environmental benefits in new technology. This policy should be implemented in the following manner:

(1) Only units constructed before 1958 should be covered.

(2) If required by the SIP, the old unit should be required to control emission levels to the extent that total emission control costs, expressed as emission control costs per megawatt hour of electricity produced, are equivalent to a new unit operating at 80 percent capacity and meeting the NSPS for that pollutant. In calculating emission control costs, it would be assumed that the old unit would operate an additional 15 years and that the "comparison" new unit would be similar and would operate for 35 years.

(3) If Item 2 does not result in compliance with ambient standards, then these old units would be allowed to meet ambient air quality standards in any way that is possible, such as through the use of tall stacks or supplementary controls. Necessary emission control requirements would then be adjusted so that the total cost for emission and ambient control is equivalent to emission control costs per megawatt hour for new units.

(4) If the above actions do not result in compliance with the appropriate SIP emission standard requirement, the old unit would be further required to reduce total atmospheric loading to an amount equivalent to a new source operating at 80 percent capacity, and meeting the corresponding NSPS by reducing the capacity factor of the old

unit. In any event, the unit must meet the ambient air quality standards.

(5) This policy can be applied to one or more pollution control requirements at the option of the old unit operator. In the absence of an election of the provisions of this policy, the regular SIP requirements should apply.

(6) Any old unit operator who has elected the provisions of this policy should not be allowed to comply with its provisions by simply complying with regular SIP requirements at any time.

(7) If an old unit operator elects to operate his unit beyond an age of 35 years, costs for the old unit and its equivalent new unit should be recalculated. The adjusted old unit cost for pollution controls over the entire period, beginning with the option to use this policy and ending with the new expected life of the unit, should be made equivalent to the adjusted cost for a new unit having the same life. This may require that additional controls be utilized by the old unit to reduce emissions to required levels.

3. Support Information: The following analysis, which was developed by the Task Force, presents the relationship between costs and effectiveness for very old unit air quality control.

In order to evaluate the costs and effectiveness of very old unit air quality controls, an analysis was done for an electric utility coal-fired power unit that was 20 years old with a remaining life of 15 years. Four different types of control were considered: (1) no control, (2) Supplementary Control System (SCS) control, (3) medium control, and (4) stringent control, Table 1. The analysis showed that control costs range from \$2.5 to \$29.7 million per year for the one plant, see Table 2. Ambient SO_2 concentrations ranged from 0.49 to 0.09 parts per million (ppm) SO_2, and annual SO_2 emissions ranged from 156,800 to 30,600 tons of SO_2, see Table 2.

Cost effectiveness was evaluated by looking at the cost per ppm SO_2 reduction and cents per pound for emissions reductions, Tables 3 and 4. The ambient controls were by far the least costly for ppm reductions, Tables 3 and 4. Also, the more stringent SO_2 emission reductions cost from 3 to 4 times more for the next increment, Tables 3 and 4.

The costs and effectiveness for the one unit were extrapolated to all of the utility units 20 years old and older based on assumptions that were made for the study, Table 5. Final results show that ambient controls with moderate SO_2 emission reductions can be accomplished for \$232 million per year (about \$1/4 billion) and that more stringent controls would cost \$913 million per year (about \$1 billion), Tables 6 and 7.

F. Air Quality Modeling

1. Policy: Standardized air quality models are required to promote equitable air quality control in the United States.

2. Rationale for and Characteristics of Models: In developing and implementing air quality control programs, it is sometimes necessary to use air quality models in order to estimate the impact of given sources on air quality. For example, air quality models must be used in the planning of new sources. In order to be equitable, air quality models must be as accurate and consistent as possible. The Environmental Protection Agency, as the national air quality regulatory agency, should be responsible for developing standardized air quality models. In developing the models, EPA should allow for sufficient public

participation through the use of public hearings and through providing opportunities for public comment. Finally, provisions should be available for the continual upgrading of standardized air quality models as more accurate information becomes available.

It is necessary that these models be applied consistently in all different types of situations. In some cases, a stringent model might be advantageous from an environmental viewpoint. In other cases, the same stringent model might be disadvantageous from the same viewpoint. However, the application of consistent models for all cases should force the development of more accurate models resulting in a more consistent public policy and the promotion of confidence in decision making.

3. Support Information: It is well known that increased coal utilization will have an impact on air quality. In many cases, this impact may be insignificant and, therefore, acceptable. During the process of planning for future increases in coal utilization, there is only one way to evaluate the impact on ambient air quality--through the utilization of air quality modeling techniques.

(1) Accurate and Consistent Air Quality Models Are Needed. The final test of the impact of a source on ambient air quality is based on the actual measured air quality once the source is in operation. The most desirable air quality models should agree as closely as possible to these measurements. It is also important that any error in the models should be equally distributed around the actual air quality data. This would make the models as realistic as possible. For example, a conservative or non-conservative model for determining the air quality impact of a specific source might be chosen, depending on the particular viewpoint of the person using the model. The

opposite model might be chosen by the same person for determining the applicability of regulations such as those involving non-deterioration of air quality. In view of this choice, the best models should be those most accurately predicting air quality with a minimal error band distributed equally around the actual data.

Consistency is another important aspect of air quality modeling. Numerous inequalities could exist around the country if different and inconsistent air quality models were to be utilized. For example, some companies might have unfair advantages over other companies if they were allowed to utilize less stringent air quality models. On the other hand, very stringent air quality models could be utilized to demonstrate that a non-deterioration policy might not be feasible in a particular area, and the use of lax or even reasonable models might indicate that a non-deterioration policy is feasible in another yet similar area. This inconsistency is not desirable. Finally, if the national policy on non-deterioration is revised, upgraded, and given more emphasis by Congress, then accurate and consistent air quality models are even more urgently needed for the reasons stated above.

(2) EPA Should Develop the Air Quality Models. EPA is the most appropriate agency to develop consistent models for nationwide use. However, in the development of these models, care should be taken to be sure that EPA does not inject biases to the point of detracting from the most accurate and consistent air quality models. This unbiased modeling can be accomplished if EPA is required to receive maximum input from the public and private sectors during the development process. Several methods of

obtaining such inputs are available. For example, EPA could hold national conferences aimed at developing the desired models from the numerous models already in use. Also, EPA could utilize such existing standardizing groups as ASTM or ASME, among others, for assistance and input in developing standard air quality models. Finally, when the necessary models are developed, they should be published and widely distributed so that they can be widely implemented and understood.

(3) Models Should be Continually Upgraded. As experience is gained with the selected models, and as more and more actual air quality data is obtained, the air quality models can be improved and made more accurate accordingly. This improvement can be accomplished by requiring EPA to periodically re-evaluate and republish information concerning the National Air Quality Models. As in the development stage, maximum input from the public and private sectors, other than from EPA, is desirable throughout this upgrading process. The periodic re-evaluation of information should occur on the order of every one to three years.

(4) Some Desirable Model Characteristics. It is not the purpose of this group to detail the characteristics of the National Air Quality Models. That should be left to modeling experts in EPA assisted by experts from other agencies. However, some suggestions can be made by the Task Force as follows:

The models chosen should be those which provide the best representation of long term averages and short term maximum concentrations. They should include provisions that permit the calculations and maximum concentrations which might occur in a variety of circumstances. Some important situations might include areas with complex terrain (with terrain heights as high as effective stack heights), and those which are dominated by sea breeze circulations.

If the models chosen use synthesized worst case meteorology rather than an objective input, then they should be accompanied by information describing the ranges of inputs appropriate for each parameter and any restrictions that the choice of one parameter may place on the others. That is, a range of internally consistent parameters should be provided. If objective input (such as one year's hourly wind speeds, wind directions, mixing heights, and atmospheric stabilities) is used, the kinds of data which are properly representative of plume dispersion should be discussed. For example, in an area with stable conditions and plume height terrain nearby, it may be that only plume height wind speeds and wind directions may be used in the analysis. In this case, a description of the models could state that plume height winds are required.

In that the aim of ambient air quality modeling is to accurately predict actual contaminant concentrations, guidelines should be established outlining the quality and quantity of ambient air quality monitoring data required for the calibration of long-term models and for the validation of short-term models.

Recommended models should have a "built-in" capability to estimate the potential error contained in predicted results. This capability should include a consideration of such factors as the quantity and quality of emission, meteorological and air quality monitoring data as well as the natural variations in geography and year-to-year variations in meteorology. Other potential sources of error should also be considered as well.

If the chosen models are complex, simpler screening models, designed to give conservative estimates, should be chosen also.

Finally, in cases where it is clear that no model at this time can be expected to give reasonably accurate results, a procedure should be established for resolving the situation via experimental techniques.

G. Emission Standards Based on Plant Output

1. Policy: Emission standards for new and existing electric and steam generating equipment should be based on gross unit output rather than heat input.

2. Rationale: Emission standards for electric and steam generating equipment are currently based on the heat content of fuel used, i.e., pounds per million Btu's.

This practice places facilities that are more efficient in converting energy fuels into more usable forms at a relative disadvantage. A power unit with a typical heat rate of 10,000 Btu per kilowatt hour (kwh) is presently permitted to emit 0.012 pounds of SO_2 per kwh, while another unit, perhaps using a combined cycle enabling a kwh to be produced using only 6,000 Btu of coal, may emit only 0.0072 pounds per kwh. The economic and social benefits resulting from a plant are primarily functions of the resulting quantity of useful product, for example, kwh of electricity and Btu of usable process heat or steam, and not the quantity of raw materials used in production. We therefore recommend that emission standards be expressed in terms of allowable emissions per unit of product for products such as kwh of electricity and Btu of usable process heat and steam. Consideration should be given to the demonstration of compliance with such standards on the basis of gross equivalent useful energy produced, design heat rate data, and the stoichiometry of the total system.

This proposal is not intended to change any of the other legal requirements in setting emission standards, such as basing NSPS on a reference system. This policy is intended to encourage improved thermal performance by giving more credit to sources that are designed to achieve higher efficiency.

H. Carbon Dioxide

1. Policy: World atmospheric carbon dioxide (CO_2) concentrations should be closely monitored, and the effects of increased levels should be projected. The Department of Energy should carry out careful studies of the sources and consequences of increased CO_2 concentration.

2. Rationale: Carbon dioxide resulting from the combustion of fossil fuels is one of very few emissions for which there is significant evidence that man's activities have produced changes in global concentrations. Specifically, there is evidence that changes of measured concentrations on the order of 5 percent have occured in the past two decades. These changes may well be the result of fossil fuel combustion.

Atmospheric CO_2 plays an important role in the earth's heat balance, and changes in its concentration may produce major impacts on the earth's climate. Continued increases in fossil fuel combustion may produce such changes. For this reason, it is important to carefully study and to project sources and consequences of increased CO_2 concentration.

The Department of Energy has begun a program to carry out these studies. This work should be done in full consultation with the scientific community, and adequate resources should be made available to implement the program.

IV. Discussion of Unresolved Issues

A. Environmental Views

Environmentalists are extremely concerned that the increased use of coal -- as proposed by the President and as supported by this Project -- will result in substantially increased air pollution in this country despite the Clean Air Act. Studies of the President's National Energy Plan show that, by the year 2000, nationwide levels of sulfur oxides will increase by 10 percent and nitrogen oxides by 70 percent from 1975 levels, even with the application of best available control technology and extensive energy conservation. (Department of Energy, Annual Environmental Analysis Report, September 1977.)

Environmentalists believe that this result is intolerable. Higher pollution levels will cause death and illness. They will cause harm to crops and other vegetation, domestic animals and wildlife, and they will interfere with the enjoyment of life. This country must continue making progress to reduce the emission of air pollutants rather than increase them. While industry has agreed that consideration should be given to amending the Clean Air Act in order to minimize increases in pollution from increased coal use, environmentalists believe that this is not sufficient. If increased coal use will produce higher emissions of pollutants, then environmentalists believe that the Clean Air Act must be amended to prevent this situation from occurring. If no other means are available, then these amendments should restrict increased coal use to prevent higher-increased emissions over present levels.

The issues that have been left unresolved by the Air Pollution Task Force were some of the most complicated problems on the agenda. In general, they were the issues that involved the widest disagreement and, once this divergence of views was recognized, the Task Force decided that our efforts would be more productive if directed to other issues where agreement was more likely.

Several of these unresolved issues were addressed by Congress in the 1977 Amendments to the Clean Air Act and neither side was willing to make a commitment to a definite position on those issues while the Congressional deliberations were continuing. Once the amendments were enacted, there was very little room left for discussion because industry, on the whole, wanted to weaken significantly the statutory language and environmentalists were largely satisfied with the outcome, although they would have liked a few strengthening amendments.

1. Significant Deterioration: From the outset of the Task Force's proceedings, this issue was a difficult one. Environmentalists prepared a lengthy paper on the implications of preventing the significant deterioration of air quality. It contained elaborate discussion of the issue and citations to numerous scientific studies documenting the serious effects on human health from pollutant levels below the national ambient air quality standards. Despite this lengthy documentation, the industry side opposed the significant deterioration concept altogether because they did not believe that significant harm to health or other kinds of injury (except for reduction of visibility) occurred at pollution levels below the national standards. They therefore did not believe that there was sufficient need to prevent deterioration of air quality as long as the national standards were not violated. Based on the industry position and because Congress was in the process of enacting provisions concerning significant deterioration, the environmentalists agreed with industry not to pursue this issue.

2. Best Available Control Technology: The Task Force's debate on this issue, like significant deterioration, was sub-

stantially mooted by enactment of the 1977 Amendments to the Clean Air Act.

The environmentalists proposed that BACT be required for all new sources, with EPA determining the appropriate system, on a case-by-case basis, and that BACT be based on the 90 percent reduction of SO_2 currently achievable. The industry proposal called for EPA to establish BACT on a regional basis subject to veto by the governor of each state and at least ten-year waivers for new and untested control systems. The 1977 Amendments basically adopted the environmentalists' position, i.e., that EPA can establish a percentage reduction of SO_2 emissions based on a BACT standard, which apparently will be 90 percent removal. Since Congress' action satisfied the environmentalists, but was substantially inconsistent with industry's position, further discussion did not appear warranted.

However, the two sides did pursue the issue of an amendment to the Clean Air Act to encourage the use of new technology for pollution control. Both sides desired to give industry strong encouragement to develop innovative control methods and we all believed that the 1977 Amendments were not adequate in this regard. The Task Force therefore devoted substantial effort to obtain agreement on this subject. This effort was successful and a policy position on new technology was adopted.

3. Application of Allowable Pollution Increments for New Pollutants: A continuing issue between environmental and industry interests is the degree of proof required to establish sufficient harm or risk from a pollutant to warrant regulatory action. This problem is exacerbated by the level of regulatory action prescribed by law once an air pollutant is named a criteria pollutant and National Ambient Air Quality Standards (NAAQS) are established in order to prevent harm to public health and welfare. The Clean Air

Act requires that the NAAQS be met within a strict deadline. In industrialized areas, this often requires a scheduled "rollback" of existing pollution, by retrofitting existing sources even though the technology may be poorly developed or non-existent for new pollutants. Because of these problems, from an environmental point of view, EPA is unacceptably slow to regulate new pollutants (or old pollutants at stricter levels) and, from an industry point of view, regulation when it does come is unacceptably harsh, rapid, and costly.

These problems could be reduced through a two-tiered regulatory scheme. The regulatory scheme proposed would establish "allowable pollution increments" (APIs) which would limit the increase in concentration of a specific pollutant in any area. The APIs would be established on the basis of substantial evidence that the pollutant caused significant harm or risk of harm without requiring irrefutable proof of such damage. If more conclusive evidence of harm is found, then the pollutant would be fully regulated in the same manner as the present six pollutants, i.e., requiring rollbacks of the level of pollutants in the ambient air to the level which, based on the evidence, is consistent with health and welfare.

This system would limit increased damage or risk of damage while more data was being gathered. It would also put pressure on industry to begin the technological innovations needed to reduce pollution since the increments would limit the size or location of new sources unless such technology was developed. At the same time, it drastically lessens the economic and social costs of regulation since no rollbacks are required of existing sources until strong evidence of harm has been obtained. Consequently, the system should be more acceptable to both environmentalists and industry than the present all-or-nothing system.

4. Air Quality Standards: The Task Force was able to reach agreement on the need for immediate review both of current air quality criteria documents and national ambient air quality standards (NAAQS). It also agreed that prompt consideration should be given to a number of other pollutants to determine whether NAAQS should be adopted for them.

The environmentalists further proposed a mechanism for dealing with situations where new air pollutants have been identified as likely to cause harm to health or other kinds of injury, but no NAAQS has yet been established. The environmentalists' proposal provided for a two-tiered regulatory scheme (see IV.A.3., Application of Allowable Pollution Increments for New Pollutants). The first tier would be "allowable pollution increments" (APIs), established on the basis of substantial evidence that a new pollutant causes significant harm or risk of harm. The APIs would only limit future increases of the new pollutant. The second tier of NAAQS procedures, and the resulting strict regulation, would require more conclusive evidence of harm.

In the environmentalists' view, this two-tier system would benefit both industry and environmentalists. It would provide for regulations to prevent substantial increases of a new pollutant when there is good reason to believe that harm is caused but the evidence of its effects is not yet conclusive. This earlier regulation would better protect public health and welfare. On the other hand, the serious economic harm resulting from rollbacks imposed as a result of adoption of an NAAQS could be delayed until more conclusive evidence of harm was obtained.

Thus, this proposal offers alternative methods of regulation. The two alternative methods allow a choice of regulation based on the amount of scientific evidence available. This makes much more sense than the present all-or-nothing approach which provides for no regulation at all or total regulation, including costly roll-backs. The environmentalists therefore believe that this proposal is worth serious consideration by Congress.

5. Non-Attainment Areas: The environmentalists and industry could not agree on a policy for non-attainment areas. Again, this was an extremely controversial issue in Congress. After Congress finished its work, the environmentalists believed that the offset provisions of the 1977 Amendments were appropriate, but required strengthening. Since industry strongly opposed such existing provisions, it did not appear worthwhile to spend further effort on this matter.

6. Old Plants: Environmentalists agree with the industrialists that the nation can better utilize its resources for reduction of air pollution by devoting large expenditures on pollution control for new and recently constructed power plants rather than expending this money on old plants (20 years old and older). As a result, environmentalists approved the "Old Plant" policy which provides that expenditures for pollution control need not be greater per megawatt hour for old plants than for newly constructed plants. If pollution control methods within the cost limits (such as coal washing and use of low sulfur coal) are not adequate to meet ambient standards, then tall stacks and intermittent controls would be permissable to ensure compliance with these standards.

The environmentalists believe, however, that these same principles should be applied to all existing power plants, whether or not they are 20 years old and whether or not they are located in non-attainment areas. Expenditures per kilowatt hour required by the Clean Air Act for newly constructed plants constitute a judgment by this country about the level of effort which is justified to control air pollution. Therefore, application should not be limited to old plants in non-attainment areas. If the environmentalists' position is adopted, then it would foster a more rational use of funds for pollution control. The money saved by application of the old plant policy would be used to accomplish greater emission reductions on more recently built plants with longer operating life. In this manner, the same amount of money can be used to accomplish more improvement in reducing overall levels of air pollutants. Unfortunately, the environmentalists were unable to reach agreement with industry about this concept of linking the relaxed old plant policy with stricter clean-up requirements for more recently constructed power plants.

However, if this principle is applied only to old plants is non-attainment areas, then it will result in a larger total amount of pollution being emitted into the atmosphere. Even if ambient standards are met in the vicinity of an old plant as a result of tall stacks and intermittent controls, these methods will cause higher pollution levels in other areas. Environmentalists believe that this additional pollution loading, even if it is below the national ambient standards, is extremely harmful because it causes harm to health, produces higher levels of sulfates, increases acid rain, reduces visibility, and causes other serious effects.

It is the opinion of the environmentalists that unless industry is compelled to increase expenditures for plants which have a greater cost-benefit ratio, it will not utilize its reduced expenditures for pollution control on old plants to obtain better control equipment for newer plants.

Environmentalists therefore support the principles in the "Old Plant" policy (see III.E.) only if it is applied to all existing plants. If this is done, all existing plants in non-attainment and attainment areas will be required to spend the same amount per kilowatt hour on pollution control as new plants. The result will be a reduction of air pollution from many plants not now subject to any emission requirements. While the environmentalists have done no calculations, they believe that these reductions will more than offset the increased pollution resulting from the application of the "Old Plant" policy to old plants in non-attainment areas.

B. Industry Views

During the process of discussion and debate, as described in previous sections of this report, the pressure of time, limitations on the extent of involvement of individuals to carry out more thorough fact finding, and the concurrent congressional action on many air pollution aspects of coal utilization resulted in several (either in part or wholly) unresolved issues. The purpose of this section of the report is to set forth, to the extent possible, the positions taken by industry members of the Task Force on these unresolved issues.

Four issues on which there was a large difference in position between environmentalists and industrialists and only limited areas of agreement are:

1. Clean Air Act Amendments of 1977/ Overall Environmental Management

2. Application of Best Available Control Technology

3. Prevention of Significant Deterioration of Air Quality

4. Siting of Plants in Non-Attainment Areas

It is crucial in discussing the industry position to acknowledge the extent to which national policy has recently been set on these issues by Congress in the Clean Air Act Amendments of 1977 (CAAA). The industry members submit, however, that in the above-named areas, the mandates of the 1977 Amendments may not establish an adequately flexible national policy to allow for the cost-effective implementation of the changes necessary to permit the increased utilization of coal while, at the same time maintaining a reasonable level of air quality improvement.

1. Clean Air Act Amendments of 1977/ Overall Environmental Management: The 1977 Amendments have given additional strength to the trend of moving air pollution regulations and policies away from national health and welfare based ambient air quality criteria as supported by a quantitative body of scientific information, toward the application of subjective judgments which support a control technology-forcing strategy as the basic regulatory mechanism. For purposes of this disucssion only the changes in Sections III (New Source Performance Standards), 127 (Prevention of Significant Deterioration), and 129 (Non-Attainment Areas) are highlighted since they relate directly to the unresolved issues dealt with by the Air Pollution Task Force.

The 1977 Amendments significantly strengthen the concept of best available control technology, which is perhaps the prime example of a technology-forcing strategy, by reinforcing existing judicial and regulatory mandates on coal burning facilities as well as by providing new legislative authority. The potential technological and economic implications of such control technology-forcing strategies are staggering for many regions of the country and must be carefully weighed against potential benefits. The potential impacts are of such magnitude, in the opinion of the industry members on the Air Pollution Task Force, that resolution of the previously named issues became virtually impossible within the time frame of the project. It would appear that all of the mandates of the 1977 Amendments have the common characteristic that rigorous scientific support is not necessary for their application through the regulatory process. This approach reflects the rationale that in the absence of a substantial body of data, health and welfare demand stringent controls regardless of costs. Such a legislative rationale significantly reduces an administrative body's initiative to technically justify its actions and renders practically ineffective the legislative call found in many parts of the 1977 Amendments to take actions "considering costs, energy, and non-air quality health and environmental impact."

Conspicuous in its absence from the 1977 Amendments is a recognition of overall environmental management, or more concisely, cross media impacts such as the impact that a reduction of air pollutant emissions will have by increasing the burden on either water or land. The lack of legislative direction in this area leads to a lack of incentive to alter the usual regulatory focus on a single pollutant discharged into a single medium.

Significant progress has been made recently in more clearly defining air pollution issues affecting health and welfare. However, rather than using this progress in such a manner that air pollution control investment can be made in the most cost effective manner, it is used in the 1977 Amendments to justify decisions which support increasingly stringent control levels across the board. The industry members of the Air Pollution Task Force strongly believe that environmental protections, where a need is indicated by significant scientific data, must be implemented. However, there must be an accountability in environmental protection in order to avoid the unnecessary application of the most costly interpretation of environmental legislation and the concomitant adverse impacts that such actions have on productivity, and other national social objectives.

The following paragraphs provide both a brief overview of the impacts of the three sections of the 1977 Amendments and a resume of the opinions of the industry members of the Task Force as to the significant factors leading to a lack of resolution on certain issues during the past year's deliberations.

2. Application of Best Available Control Technology: Section III of the Clean Air Act on New Source Performance Standards (NSPS) for fossil fuel stationary sources is to be revised within the coming year to include an allowable emission rate and a percentage reduction requirement that reflect the best technological system of continuous emission reduction, "considering cost, energy, and non-air quality health and environmental impact," which EPA determines has been adequately demonstrated. Credit can be given for precombustion beneficiation processes, and a variable percentage reduction may be considered by EPA.

This provision will no doubt result in more stringent NSPS for SO_x and NO_x, and particulate emissions. The revision implies the use of flue gas desulfurization, combustion NO_x control with resulting adverse operating impacts on efficiency, maintenance, and plant availability and reliability (increasing the need for more coal generating capacity or at least peaking capacity), as well as the increased use of baghouses for fuels from which the ash is not effectively collected in electrostatic precipitators. Chemical coal cleaning would be effectively removed as a sole means of achieving NSPS, thereby eliminating a new technology that appears to offer significant advantages over alternatives in many cases. Sources may request waivers from NSPS requirements to use innovative technology. However, the conditions are of such a restrictive nature that the provision serves essentially no use as an incentive (see policy recommendations on New Technology, III.C.).

It has been the industry position that performance standards for air pollutant emissions must be directly related to the impact that those emissions have on ambient air quality. A system of emission limitations is *directly* related to the control of ambient air quality and, in the opinion of the Task Forces' industry members, it is the most logical approach to improvement and maintenance of air quality. It is recognized, however, that the quality of fuel has a direct impact on the amount of control required to achieve specific emission limitations and, therefore, certain areas of the country may have to utilize more stringent control procedures than others. Additionally, certain areas of the country have environmental quality problems that differ from other areas, and thereby require a more stringent level of control. An emission limitation control approach alone, as set forth in the 1970 Amendments, allowed some flexibility for states to establish more stringent controls where dictated by local conditions. On the other hand, the 1977 Amendments eliminate this flexibility and impose the most stringent control on every part of the country without due regard to local need. Industry firmly believes that there are many regional differences which will affect the choice of control levels when a best available control technology (BACT) approach is utilized as required by the 1977 Amendments. The broad application of BACT without due regard for the *need* for the most stringent level of control will likely cause unnecessary and wasteful expenditures in order to achieve the projected levels of coal use that will be required to attain national energy independence goals. The industry members of the Task Force strongly believe that the mandates regarding BACT in the 1977 Amendments cannot be justified as being necessary to the protection of public health and welfare, and must be modified in order to allow for more cost effective methods of improving air quality.

3. Prevention of Significant Deterioration: Section 127 of the 1977 Amendments,

concerning the prevention of significant deterioration (PSD) of air quality, has for its purposes: (a) the protection of public health and welfare from any actual or potential adverse effect which the Administrator determined may reasonably be anticipated to result from air pollution "notwithstanding attainment and maintenance of all national ambient air quality standards," (b) the preservation, protection, and enhancement of air quality in areas of special national or regional recreational, scenic or historic value, (c) the insurance of consistency between economic growth and existing clean air resources, (d) the assurance that a source in one state will not contribute to significant deterioration of air quality in any other state, and (e) the assurance that decisions regarding increased air pollution to which this section applies are based on an evaluation of all the consequences of such a decision and that public participation is included in the decision-making process.

The Act mandates a three class system. Mandatory Class I (Pristine) areas have been defined as all international parks, existing national wilderness areas greater than 5000 acres, national memorial parks greater than 5000 acres, and national parks greater than 6000 acres. All other areas are initially classified as Class II. Indian reservations have exclusive authority to redesignate tribal lands as any class. The maximum allowable increase over baseline concentrations of SO_2 and particulates will be, respectively, two (annual arithmetic mean) and five (annual) geometric mean) ug/m^3. In all cases, new power plant applicants must demonstrate that BACT will be used. This will be determined on a case-by-case basis, with NSPS essentially forming the lower boundary. In addition, EPA must promulgate PSD regulations for NO_x, hydrocarbons, CO, and oxidants within two years.

The stringency of these mandates may for all intents and purposes (when combined with the limitation of stack heights in the Amendments) preclude the construction of large new coal burning facilities. The BACT provisions of this section may be interpreted, in cases where plants are forced into non-attainment regions, to require combinations of control steps far exceeding NSPS requirements (such as low sulfur coal, coal cleaning, and flue gas desulfurization). The industry members of the Task Force strongly believe that the PSD mandates in the 1977 Amendments go far beyond that which is necessary to accomplish the purposes of section 127 in a cost effective manner. In numbers sufficient to cause national concern, fuel burning industries, particularly the utility industry, may be forced to forego well-demonstrated economics of scale for central power stations because of the need to build several scattered small coal fired power plants in place of a large one. In addition, other environmental costs, believed to be unnecessary by the industry members, would be imposed due to duplicative fuel handling, transportation systems, and transmission facilities. An optimally-sized (both economically and environmentally) plant presently consists of several units from 500 to 1000 megawatts each, for a total plant capacity of 3000 megawatts or higher. Admittedly, cogeneration may lead to the creation of some smaller generating facilities at selected sites, or may in some cases even decrease utility capacity needs for industrial customers. Practically speaking, however, such facilities will not be the rule for several decades.

As previously indicated, the industry members strongly recommend that accountability in environmental management be implemented. The costs associated with PSD, in addition to the older requirements of clean air mandates, would be over and above those costs imposed on the public for the compliance of other industries with environmental requirements. Sufficient attention has not been given to providing the general public with an opportunity to make a value judgment as to whether the additional costs of compliance with PSD mandates are justified (considering that PSD is over and above that which is necessary to control any major public health problem). Financial resources that will be directed to PSD mandates will not be available for other social needs. This is not to say that these environmental expenditures may not be valuable, but rather, that the incremental costs associated with PSD should be consciously balanced against the relative importance of other social needs.

The national average costs as quoted during the congressional hearings on the 1977 Amendments, do not adequately represent the individual hardships that the PSD mandate will be likely to impose on individual citizens in different areas of the country. The industry members of the Task Force believe that detailed individual analyses either for separate utility systems or for different regions of the country could best identify potential hardships, and that such studies must be carried out in order to provide accountability in environmental management. The PSD provisions have the effect of imposing an environmental mandate that overrides other needs of our society such as housing, jobs, and health care, to mention only a few. This situation exists because the costs of the PSD provisions have not been adequately balanced against these other needs.

4. Siting of Plants in Non-Attainment Areas: Section 129 of the 1977 Amendments in essence allows EPA to waive, before July 1, 1979, source-by-source emission offset requirements if all of an applicant's facilities within a particular state are in compliance with the Clean Air Act and new sources achieve the "lowest achievable emission rate." This would be the lower of the most restrictive emission limitation in any state implementation plan for a particular source (unless the applicant can demonstrate that the limit

is not achievable) or the lowest emission level achieved in practice by any such source. This may very well cause an automatic ratcheting provision for control requirements on all criteria pollutants in the near term, even though only some such pollutants for instance particulates and oxidants, exceed ambient standards in many regions of the country. The industrial members of the Task Force believe that the mandate of the 1977 Amendments in combination with administrative practices recently established by EPA will have serious and in many cases devastating effects on many of the nation's industries. It is believed that many older and less efficient plants of marginal compliance may be kept in service in order to preserve perceived "air rights" when they may very well otherwise have been replaced by cleaner, new plants. In other cases, non-attainment mandates in the 1977 Amendments may have severe impacts that are potentially devastating on local economies. Such impacts could limit expansion, force the closing of marginal installations with associated loss of jobs without providing adequate replacement jobs in areas where they are or would be needed, and further reduce a dwindling tax base in some areas, so as to adversely impact various local educational, welfare, and other social programs. Industry believes that in order to allow the increased utilization of coal in a cost effective and accountable manner, more flexibility must be put into the national policy on expansion in non-attainment areas. This flexibility would allow for a limited number of variances from some of the technical requirements of the existing laws and policies for compelling reasons. In all circumstances, eventual compliance with ambient standards would be required.

The industry members of the Air Pollution Task Force have summarized above several of the significant controversial issues that were discussed extensively but remained unresolved during the past year of deliberation. Although, in our opinion, these issues are significant to the national efforts to accommodate increased utilization of the country's coal resources in an environmentally acceptable manner, considering not only costs and benefits but also the achievement of other social objectives, their irresolution by no means detracts from the good rapport and improved understanding of each side's positions that has resulted from the past year's deliberations. The degree of agreement that was achieved (as exemplified by the Policy Recommendations, Section III), in the opinion of the industry members, was well worth the effort put forth, and it appears reasonable to assume that further progress could be made through more extensive deliberations on the unresolved issues in this last section of the report.

List of Acronyms and Abbreviations

AIR POLLUTION TASK FORCE

BACT - Best Available Control Technology

EPA - Environmental Protection Agency

NAAQS - National Ambient Air Quality Standards

NSPS - New Source Performance Standards

SIP - State Implementation Plan

SCS - Supplementary Control System

TSP - Total Suspended Particulates

Glossary

AIR POLLUTION TASK FORCE

air quality model - a mathematical relation for predicting air quality concentration, in terms of mass/unit volume, based on input parameters such as plant operating and atmospheric conditions.

atmospheric loading - the total amount of pollution present in the atmosphere; expressed in terms of mass.

emissions - the quantity of pollution emitted at a given point; expressed in terms of mass/unit time.

emissions standard - a standard restricting the emissions of a given air pollutant; expressed in terms of mass/unit time, or mass/unit heat input, or plant output.

gross plant output - gross output from a given plant; eg., megawatt hours of electricity.

heat input - expressed in millions of Btu's; the amount of energy input into a plant; eg., burning of fossil fuel.

particulate - a solid or liquid particle air pollutant.

plume height - height of the center line of a plume over a given ground point.

plume - a visible air pollutant effluent; usually emitted from a stack.

pollutant - from the standpoint of abatement, two

basic types of pollutants are recognized - non-degradable and degradable.

pollutant loading - total amount of pollution in the atmosphere at any given time.

proven technology - a device for controlling atmospheric emissions that has been proven technologically feasible.

respirable fraction - the particulate air pollutants that are inhaled into the lower respiratory tract and are usually considered to be less than ten micrometers in size.

source - a facility that emits air pollution as a part of its process.

stoichiometry - the branch of chemistry treating laws of chemical combination and of the relations between the properties of substances and their composition.

supplementary controls - systems designed to intermittently control air pollution concentrations at ground level based on atmospheric and plant operating conditions.

tall stacks - devices used to emit and disperse air pollutants at a level above ground; usually 500 feet or higher.

SECTION 3
FUEL UTILIZATION AND CONSERVATION TASK FORCE REPORT

TASK FORCE MEMBERS:

Paul P. Craig,
Co-Chairman

Matthew Gould,
Co-Chairman

Clark Bullard

Norman L. Dean, Jr.
Frank G. Feeley
John H. Gibbons
Jerome Kohl
Herbert D. Nash
Robert Williams

ADVISORS:

Byron R. Brown
James O'Connor
William B. Wilson

TECHNICAL STAFF:

Susan J. Finnegan

TABLE OF CONTENTS

REPORT OF THE FUEL UTILIZATION
AND
CONSERVATION TASK FORCE

I. Introduction

A. Objectives

The purpose of the National Coal Policy Project was to promote discussion and interaction between environmental and industry spokespersons with the hope of increasing understanding of the issues involved in the changing of our patterns of coal utilization. The objective of the discussions between the two groups was to identify areas of consensus for rational coal utilization. Areas of disagreement or areas where consensus did not occur were to be identified also.

The rapport established between the industry and environmental members was viewed as a major accomplishment of the Task Force. Despite the existence of many areas of substantive disagreement, all participants were pleased to discover the ease with which productive discussions could be held.

A broad range of subjects related to coal utilization and coal technologies were covered. Even where consensus was not achieved, the discussion of issues did serve to focus thinking and to identify areas of specific disagreement. Disagreement tended to occur both within and between the industry and environmental groups. Many of the opinions as to the economic and technical feasibility of specific coal utilizing technology did not align with preconceived notions of industry or environmental positions, but tended to be derived from individual backgrounds independent of group identity.

The Task Force found that agreement on general principles was far easier to achieve than agreement on specifics. The closer the discussion came to specifics, the more divergence occurred. In large measure, this stemmed from the need for increasingly precise information as specific recommendations were approached. Policy statements, of the sort developed by the Task Force and presented here, formed the foundation for specific agreements. It is recommended that the next phase of this Project, or the focus of a succeeding project, be upon specific studies, with an emphasis upon progressing from general principles to explicit recommendations.

B. Common Themes

There were several areas of mutual agreement upon which specific recommendations were developed. These themes are as follows:

1. A belief that energy conservation must be a cornerstone of any rational national energy policy.

2. A belief that some expansion of the use of coal in the United States is a necessary component of a rational national energy policy.

3. A belief that the use of coal should be viewed as an intermediate step toward the use of long-term, sustainable resources.

4. A belief that strategies which minimize governmental involvement (e.g., natural gas deregulation) are to be preferred over those which encourage expansion of such involvement. (Governmental intervention is, of course, clearly called for in certain situations, e.g., for pollution control and in situations in which natural monopolies exist.)

The reasons behind the belief in these common themes varied considerably, but all participants felt that coal leaves much to be desired as a fuel source. Coal reserves are large, but finite. Within a century or so, coal will have to be replaced by self-sustaining resources. Since it will require many decades to make a transition to self-sustaining fuels, this process should be started immediately. The urgency of such a transition may become much greater should long-term climatic changes associated with carbon dioxide or other global effects begin to be recognized. While all participants are aware of these problems, environmentalists were considerably less at ease with the need for coal expansion than were the industrial participants since the use of coal presents so many environmental problems. In any case, these differences in justification did not detract from the importance of the consensus on the themes listed above.

C. Future Use of Coal

It is a presumption of this study that the use of coal is likely to increase during the coming decades. This increase will occur regardless of the evolution of the remainder of the energy system. A number of studies and policy documents have attempted to project coal production in 1985. Among these, we note the following:

1. The National Energy Plan (Executive Office of the President, April 29, 1977) estimates production in 1985 of 1070 million tons without the plan, and 1280 million tons with it. Reservations about these estimates were expressed by many groups, including the Congressional Budget Office.1/

2. ERDA estimates various production levels in the range of 863 million tons to 1066 million tons.2/

3. The Workshop on Alternative Energy Strategies (WAES) estimates the production capability of the U.S. between 726 and 944 million tons in 1985.3/

4. The Office of Energy Programs of the Domestic and International Business Administration, U.S. Department of Commerce,

---

1/ Congressional Budget Office, Staff Working Paper (June 1977).

2/ Energy Research and Development Administration, 48 (June 28, 1975).

3/ The Workshop on Alternative Energy Strategies, Energy: Global Prospects 1985-2000 (McGraw Hill Book Company, 1977).

estimates production of coal in 1985 at 776 million tons.4/

These estimates may be contrasted with 1975 production of 665 million tons of bituminous coal and lignite, and 6.2 million tons of anthracite.5/

The Task Force has not evaluated these projections and, moreover, believes that it is extremely difficult to assess future demands. It has no opinion on which of them is best. The extent of expansion of coal use will therefore depend upon a number of factors, of which the following are of primary importance:

1. The price of coal relative to the prices of other inputs to production, such as other fuels, labor, and capital.

2. The rate of the application of coal utilizing techniques, which is in turn affected by economic and environmental factors.

3. Energy conservation activities, such as the efficiency of coal use.

4. The rate of growth of national demand for energy, as affected by economic growth, energy conservation, public policy, etc.

5. Availability of competitive alternative energy forms, for example, petroleum, natural gas, nuclear power, solar energy, etc.

4/ U.S. Department of Commerce, Forecast of Likely U.S. Energy Supply Demand Balances for 1985 and 2000 and Implications for U.S. Energy Policy (Office of Energy Programs, Domestic and International Business Administration, January 20,1977).

5/ Congressional Research Service, Energy Information Handbook, report for the Subcommittee on Energy and Power of the Committee on Interstate and Foreign Commerce, U.S. Congress, House of Representatives (July 1977).

6. Progress in commercialization of synthetic fuels based upon coal.

7. The process of creation and implementation of public policy, such as national security, environmental and safety policies, and the timing of these policies.

II. Principles Guiding the Rational Use of Coal

The Task Force reached consensus on principles in three areas including social costs, energy pricing, and environmental stewardship.

A. Social Costs

Recommendation

Members of the Task Force are in agreement as to the desirability of identifying and internalizing relevant social costs, e.g., environmental and safety costs.

Discussion

The Task Force found that social costs were viewed differently by industry and environmentalists. The following discussion endeavors to express how the decision-making process of the two groups differs with regard to these costs.

From a business point of view, energy related decisions should be made on the basis of the best estimates of capital and operating costs of facilities and availability of fuels over the lifetime of the facility. Once assumptions are made as to what these costs may be, then conventional economic computations (e.g., return on investment) can be used to arrive at a decision on which investment route to follow. These decisions are always tempered by an assessment of future uncertainty, and decisions are often strongly influenced by considerations of diversity of fuel supply, reliability of supply, etc. In each case, however, the investment decision is made on the basis of the impact of that decision upon the company, for the responsibility of a corporate decision-maker is to the company owners.

On the other hand, environmentalists view the decision-making process in a broader sense, and are concerned that decisions on energy mix be made on the basis of societal costs. Decisions made on these two bases often differ.

In discussing the above viewpoints, the Task Force found that the two points of view may not be irreconcilable. Industry spokespersons claimed that if social costs were made known or were quantified through regulatory or other legislation, then they would be taken into account in the decision-making process. Thus, the problem is to identify the costs and to internalize them through legislation. However, it was recognized that dollar costs cannot be assigned in many cases.

Environmentalists believe that it is the role of the government to identify the social costs associated with environmental pollutants and to establish standards based upon an assessment of these costs. Such an assessment should take into account all significant factors, including those difficult or impossible to quantify.

B. Energy Pricing

Recommendation

The Task Force recognizes that there are substitutes for energy in providing goods and services. The preferred common bases for trade-off decisions between supply and conservation alternatives should be minimum total costs, including social costs. Consequently, whether it be a supply or a conservation action, price should reflect total cost, including social cost.

Discussion

These are long-term goals as practical difficulties constrain rapid progress toward them. However, the recommendations of the Pricing Task Force represent a step in that direction. In the short-term, approaches such as standards, codes or regulations may properly be used as means for approaching these goals.

C. Environmental Stewardship

Recommendation

Government established air and water pollution control strategies are the best means to internalize the societal costs of pollution.

Discussion

Members of the Task Force agree that government regulations concerning pollution abatement directly affect the timing and potential use of coal-utilizing technologies, since all new coal utilizing systems must meet EPA and state pollution standards. A major motivation for development of many new technologies (e.g., fluidized beds and low Btu synfuels) is to burn coal cleanly. Widespread use of new technologies will depend on their ability to achieve this goal economically.

D. Energy Conservation

The Task Force believes that a ton of coal "saved" is worth more to the nation than a ton of coal mined.

Therefore, conservation should be given the highest possible priority in government energy programs. While it is true that government's attention to conservation has been greatly heightened over the past several years, it is clear that conservation still ranks far behind supply in government priorities.

1. Benefits of Conservation: Energy conservation offers several important advantages over increased supply. Conservation generally results in less damage to the environment than supply expansion. Conservation often requires a smaller front-end investment than does supply expansion. Also, the overwhelming majority of evidence indicates that conservation actions not only save energy, but can also be profitable (see the Appendix at the end of this report for examples).

2. Barriers to Conservation: Despite the significant benefits of conservation,

it continues to receive much less attention than the Task Force feels it is due. The Task Force identified several reasons for the lack of impetus in the area of energy conservation:

a. Higher rates of return are sometimes required by the consumer for energy conservation investments than generally would be obtained for other investments. This situation appears to occur not only in the residential sector but also in medium- or small-sized industries or plants. (In some large industries, especially energy intensive industries, energy conservation investments are treated preferentially.) Reluctance to commit capital to non-product related projects can lead to a slower implementation of conservation than might be in the best interest of the nation.

b. Capital for energy conservation investments is not readily available: many firms, especially small- and medium-sized operations, have an insufficient ability to raise the capital necessary for such investments.

c. Insufficient information exists on energy conservation opportunities: the small- and medium-sized firms often do not have sufficient information as to how and where they are consuming energy, the technical opportunities for conservation, and the true costs and benefits of those opportunities to the firm. This is less of a problem in energy intensive industries where energy costs are a significant percentage of basic manufacturing costs.

d. Lack of management commitment: in some instances, industrial conservation has been hindered by a lack of commitment on the part of industrial management and employees to reducing energy consumption.

e. Government laws and regulations may present barriers: in some instances, government laws or regulations present a barrier to conservation. Examples of such laws include those prohibiting industry from selling cogenerated electric power, and those prohibiting the turn-key construction of public resource recovery projects. The Task Force recommends that every effort be made to reform laws and regulations that present barriers to the efficient use of energy.

f. Regulated utilities have little economic incentive to encourage their customers to conserve energy, or to cooperate with their customers in energy conservation projects such as cogeneration.

g. The cost to any given industry of obtaining new energy supplies from a utility is often less than the cost to the utility of obtaining new energy supply capability.

Recommendations

To help alleviate these barriers and to affirmatively encourage conservation, the Task Force makes the following recommendations:

1. The use of natural gas as the primary fuel in large steam generating plants should be phased out except when required by environmental considerations. Exceptions should be permitted for small units.*

---

* The industrial members of the Task Force believe that exemptions should be permitted for units with energy inputs of less than 50 million Btu/hr. The environmental members felt it inappropriate to include a specific number without carrying out a full-scale analysis of the justifications for such a number. However, available time did not permit such anlysis.

2. The use of petroleum as the primary fuel for the generation of steam in large units should be phased out.* Exemptions should be permitted for smaller units.

3. Energy conservation technology transfer services should be rapidly expanded and funding for such services should be substantially increased.

The government should expand its efforts to disseminate information on both coal use technology and energy efficient technology. Such efforts should be made at the state level to the maximum extent possible, and should be directed mainly at small- and medium-sized energy consumers, since the smaller users are those least likely to have the informational resources to keep abreast of the latest technology.

In particular, it is the opinion of the Task Force that the Department of Energy's (DOE) Energy Extension Service concept should be expanded as quickly as possible from its present involvement in 10 states to include each state that desires to establish an extension service. Such state extension services should aim to improve market signals, provide training in energy technology and efficiency, and provide direct technical and economic assistance to energy consumers. The extension services should work with the federal laboratories and regional offices of DOE. To the maximum extent possible, they should make use of already existing technology transfer services such as those provided by universities, agricultural extension services, and private consulting and service organizations.

---

* The industrial members of the Task Force believe that exemptions from this recommendation should be permitted on a case-by-case basis for units with energy inputs less than 300 million Btu/hr. Environmentalist members felt it inappropriate to include a specific number without carrying out a full-scale analysis of the justification for such a number. However, available time did not permit such analysis.

4. Limited use of tax incentives. The most desirable mechanism for encouraging investment in energy conservation is free market pricing.

A general tax incentive program for energy efficiency investments is unworkable. It is impossible to define eligible investments without excluding many of the most important projects, inhibiting innovation, or including many projects that would have been undertaken without the tax incentive. Therefore, tax incentives, such as liberalized investment tax credits, should be used only for very specific high priority investments such as cogeneration, solar heating, and coal conversion.

III. Near Term Coal Utilization Policy

Near term coal policy should be designed so as to encourage the use of coal to replace oil and gas. Replacement strategies should always aim towards the minimization of environmental impacts. The key opportunities for near term action are in the areas of conversion of existing facilities and cogeneration.

A. Conversion of Existing Facilities

1. Coal conversion programs should be based upon a realistic appraisal of the availability of sufficient coal.

2. Coal conversion programs must conform to environmental standards, especially those associated with state air pollution plans.

3. Small units should be exempt from mandatory conversion requirements.*

---

* The industrial members of the Task Force believe that exemptions from mandatory conversion should be permitted for units with energy inputs below 100 million Btu/hr. Environmentalist members felt it inappropriate to include a specific number without carrying out a full-scale analysis of the justification for such a number. However, available time did not permit such analysis.

4. The regulatory environment for coal conversion should, insofar as possible, be stable so as to encourage effective decision-making processes that are unhindered by frequent changes in the regulatory requirements.

5. The conversion of energy systems which did not previously burn coal is likely to be economically infeasible in most cases. Case-by-case studies must be made based upon present and projected costs of competing fuels, and upon the total cost of conversion. Age and costs will affect the feasibility of conversion.

While conversion of gas and oil-fired boilers to coal is desirable, it is often difficult to accomplish (and may be impossible) for units that were not initially designed to fire that fuel. For industrial size units (fifty million to one billion Btu/hr), coal burning boilers and necessary support facilities far exceed oil and gas burning units in cost. In larger utility size units, the differential is not as great. However, an added cost allowance must be made for pollution control equipment and the cost escalation during construction that will inevitably occur since coal-fired units take much longer to design and build.

Conversion is not a simple matter of removing existing apparatus and substituting coal-burning equipment in its place. There are many instances where it would be extremely difficult, very expensive, or even physically impossible to replace existing oil or gas boilers with coal-burning facilities. The attached figures illustrate this point.

The small silhouette (Figure 1) is an oil-fired boiler recently installed at a plant located in the south. At the same location and adjacent to the oil-fired boiler, a coal-fired boiler was installed as shown in Figure 2. This silhouette includes the bunker, the boiler, the baghouse installed to meet particulate emission control standards, and the interconnecting ductwork. Good storage facilities are not included. In this case, the top of the coal bunker is approximately 75 feet above grade. Both units have the same capacity of approximately 190 million Btu/hr. Obviously, the

Figure I. Oil-Fired Unit

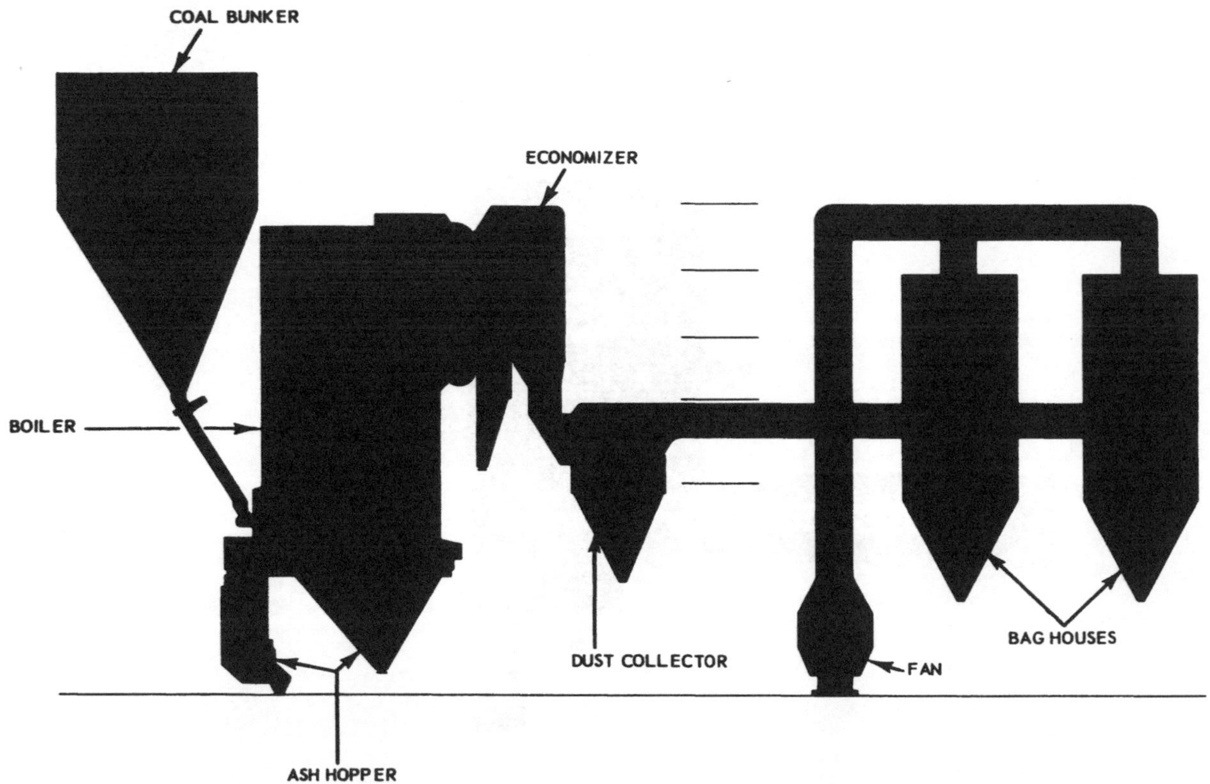

Figure II. Coal-Fired Unit

oil-fired package unit is not "convertible" to coal. Substitution of coal for oil would require complete replacement of this unit, a capital intensive and time-consuming procedure. If manufacturing and process equipment are built around the power source, which for energy efficiency purposes is often the case, space limitations might prevent substitution of a coal-fired boiler for an existing oil- or gas-fired unit.

B. Cogeneration

Cogeneration is defined as that process in which after raw energy resources are converted into mechanical or electrical energy through the use of turbines, internal combustion engines, or other heat engines, the by-product heat and system exhaust is used to provide a source of energy for a manufacturing or processing operation, including space heating. Cogeneration may also be defined as the converse of the process described above (e.g., bottoming cycles). The cogeneration or cooling system may serve one or more users and may be serviced by the industrial or other heat or power user, the utility, or a "third party."

Findings

1. Current higher energy prices and the anticipation of even higher future prices have reversed the historically ebbing United States interest in cogeneration. The upward trend in energy prices coupled with advances in the state-of-the-art of combustion/cogeneration technology and a major program to shift industrial and utility boilers to coal, all provide strong incentives to reverse this trend away from cogeneration.

2. Cogeneration technology, while capable of further advancement, is presently well-proven, economically viable, extensively used, and is saving considerable energy. There is room for much expansion of cogeneration with considerably less lead time than is required for large new power plants.

3. While it is technically possible to implement a great deal of cogeneration by 1985, institutional constraints and barriers will considerably limit the extent to which such implementation is likely to be achieved. Barriers and constraints to cogeneration and coal cogeneration include the following:

a. Electric utility, federal and state utility, and other regulatory policies (including low-buy and high-sell pricing rates, control of environmental emissions, and possible imposition of regulatory controls on a new cogenerating industry). This affects the prices utilities offer for cogenerated electricity.

b. Internal concerns and the reluctance of industry to implement cogeneration (e.g., the extra operations and maintenance costs associated with a cogenerating plant, load characteristics, lack of capital, a reluctance to engage in an activity foreign to their main business, or a new type of activity).

c. Uncertainties about future energy supply and prices, buy/sell arrangements, and standby charges.

4. Cogeneration will probably be implemented at considerably different rates in various parts of the United States.

5. The implementation of cogeneration will result in increased fuel consumption within the industrial sector (excluding electric utilities), but decreased total energy demand (including electric utilities). In the near term, electricity produced by back pressure steam turbine cogeneration will be used mostly by the industry that produced it, but, if in the long-term diesel and combustion turbines are increasingly used for cogeneration, considerable surplus power can be produced for sale to others.

6. Vigorous public support of the implementation of cogeneration over the next several decades could save oil equivalents, variously estimated, in the range of 680,000 bbl/day by 1985 to 2.5 million bbl/day by 2000.* Public support of the removal of present regulatory barriers, as well as tax and regulatory incentives, is justified because of national interest in fuel conservation.

7. The typical size of cogeneration units is considerably smaller than a modern power plant, thereby enhancing the opportunity for decentralized energy systems.

8. The best payout is obtained when cogeneration facilities are installed as new equipment -- as compared with payout for the replacement of operating equipment that had a remaining operating life.

---

* Resource Planning Associates has estimated the impact of various governmental actions on installed cogeneration at factors of 2 to 3 depending on the actions. See Resource Planning Associates Interim Report of June 17, 1977, "The Potential for Cogeneration Development in the Chemical Industry by 1985," prepared for FEA, Office of Conservation.

Dr. Robert Williams (of Princeton Center for Environmental Studies) made a study which concluded that 200 GW (e) of central station equivalent cogeneration capacity (energy savings of 2.5 million bbls/D oil equivalent) is a reasonable target for the year 2000. See R. Socolow, "The Coming Age of Conservation," Annual Review of Energy, Vol. II (1977).

A Dow Chemical study for the National Science Foundation, Office of Energy R & D Policy found that the equivalent of 680,000 bbls/day could be saved by 1985 using steam turbine topping cycles if all industrial installations generating over 400,000 pounds of steam per hour cogenerated.

9. Combustion turbine systems can be purchased for oil or gas firing. However, it is desirable that the purchasing of some oil-fired cogeneration systems be encouraged in order that experience in operating these systems can be gained. Development work is going on with turbines that use coal in an oil slurry or in closed (or indirect) cycles, some of which utilize fluidized beds.

10. A premature large scale switch to coal could preclude the use of cogeneration. If government policy encourages the rapid conversion of industrial boilers to coal, especially for larger sized units, then the major potential of cogeneration in the next few years will be preempted.

Recommendations

1. Department of Energy coal conversion and reconversion rules should be reformulated to:

   a. Promote the incorporation of a cogeneration option.

   b. Pace the conversion program to take maximum advantage of advanced technologies, such as fluidized bed combustors that could facilitate cogeneration and provide multi-fuel capability (e.g., municipal wastes, coal) and minimize environmental emission problems.

2. State and federal policy initiatives should include:

   a. Provisions for arbitration by the state utility commissions in the event that an industry cannot negotiate a satsifactory contract with a utility and requests the intervention of the utility commission.

   b. Procedures to exempt qualifying facilities from the Federal Power

Act, the Public Utility Holding Company Act, and where appropriate, from state utility regulations.

c. Provisions for the federal government to provide technical assistance and research and development to facilitate cogeneration.

3. EPA New Source Performance Standards relate allowable emissions to Btu input to the boiler. A more appropriate standard would be geared to the useful energy output rather than input. Currently, cogenerating facilities are being penalized under NSPS. (See the Report of the Air Pollution Task Force for further details.)

4. Actions to encourage cogeneration growth that are consistent with long-term replacement costs should be required of federal agencies (e.g., TVA, BPA, DOD). Other public bodies should be provided with federal direction and incentives, including technical help, to encourage the installation of cogeneration facilities.

5. Personnel training programs relevant to cogeneration should be developed.

IV. Government Involvement in Research, Development, and Commercial Stages of New Technologies

A number of new technologies for the use of coal are presently being developed by industry, the federal government, and other nations. The Task Force believes that the development of new coal technologies is of the highest national priority. These technologies offer the prospect of new ways to use coal at lower cost, with reduced environmental impact. At the outset, the Task Force had hoped to be able to make specific recommendations with respect to new coal technologies but this proved to be impossible. Each emerging technology had its advocates, and in most cases the arguments for or against a particular technology were highly complex and often involved proprietary information.

Recommendations

1. The government has a role to play in promoting technological research and development. Basic and applied research should be encouraged and funded by the government. As new technologies are developed, the role of the private sector should increase.

2. Members of the Task Force feel that financing the commercialization of new technologies is not primarily the responsibility of the government. Plants for commercial use can best be funded through private enterprise.

3. In determining which technologies to develop or encourage, both economic and environmental objectives should be considered. A potentially cheap technology should be encouraged only if there are good prospects for environmental control.

A. General Discussion

The discussion of new technologies centered around economic vs. environmental concerns and criteria. Environmentalists generally felt that in deciding which technologies should be developed, the prime criteria should be environmental impacts. Technologies with great promise for decreasing pollution, but which are as yet economically unsound, should occasionally still be pursued. Industry, on the other hand, felt that economic concerns should be the prime criteria for development. In choosing technologies to develop, industry would first choose those which are the most economically attractive, and then try to develop them to operate within environmental standards.

These two viewpoints show a difference in approach which may prove crucial in determining the selection and course of new technology.

B. Alternative Ways to Burn Coal Cleanly

The Task Force felt it useful to include a summary of the alternative ways of utilizing coal in a clean manner. The tables following this section summarize processes currently being explored in this regard. This summary is not meant to cover all the possible issues involved in the use of different processes, but only to summarize the alternatives. Specific issues are too numerous to be covered in a policy project of this scope.

V. Unresolved Issue: The Role of Electricity

The Task Force discussed at some length the question of whether emphasis on technologies that produce electricity from coal should be favored over those that produce other products (e.g., synthetic gas or oil). No consensus was reached on this issue. Arguments for electrification were presented and, in turn, countered with rationales for gasification and liquefaction technologies. Neither industry persons nor environmentalists were convinced of the merits of either approach. The choice of which route to follow should be based on individual circumstances, which may vary from location to location. Accordingly, no recommendations are offered by the Task Force.

FIGURE 3. CLEAN FUELS FROM COAL

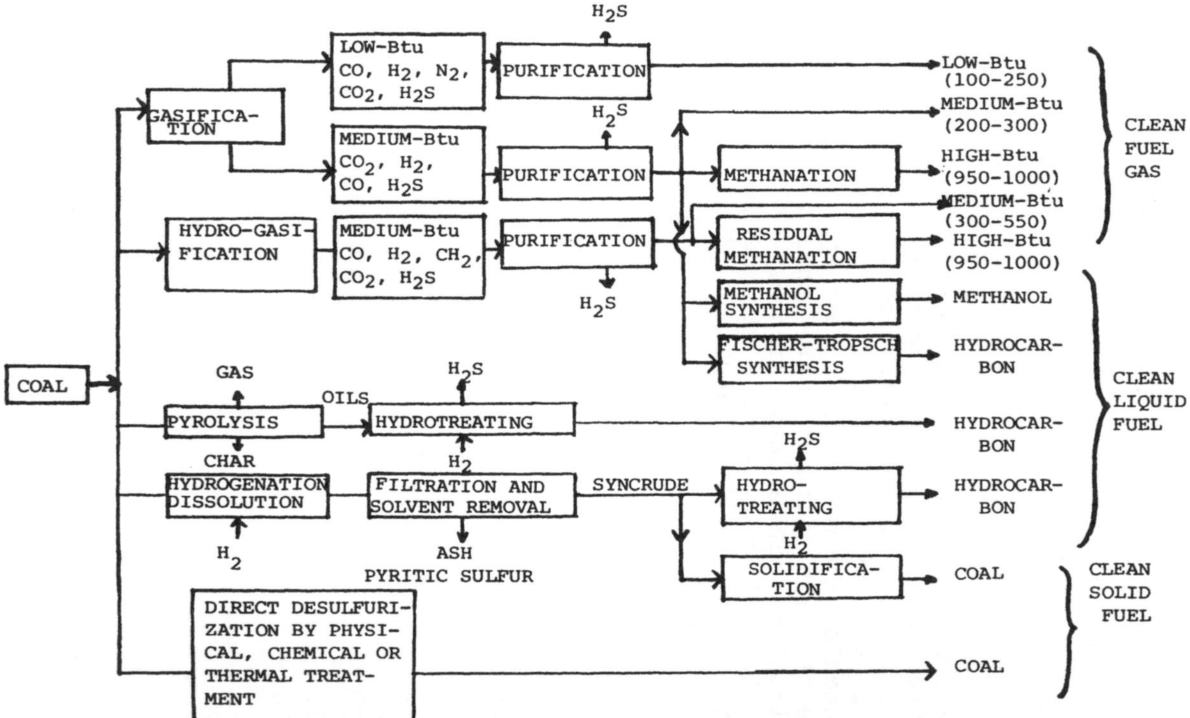

SOURCE: <u>Clean Fuels From Coal Symposium 2 Papers</u>, Institute of Gas Technology, Chicago, Illinois, October 1975. Date of Symposium--June 23-27, 1975, "Clean Fuels From Coal" by William Bodle, K.C. Vyas, T. Talewalkar.

figure 4
COAL CONVERSION PROCESSES AS APPLIED TO MARKET SECTORS (5)

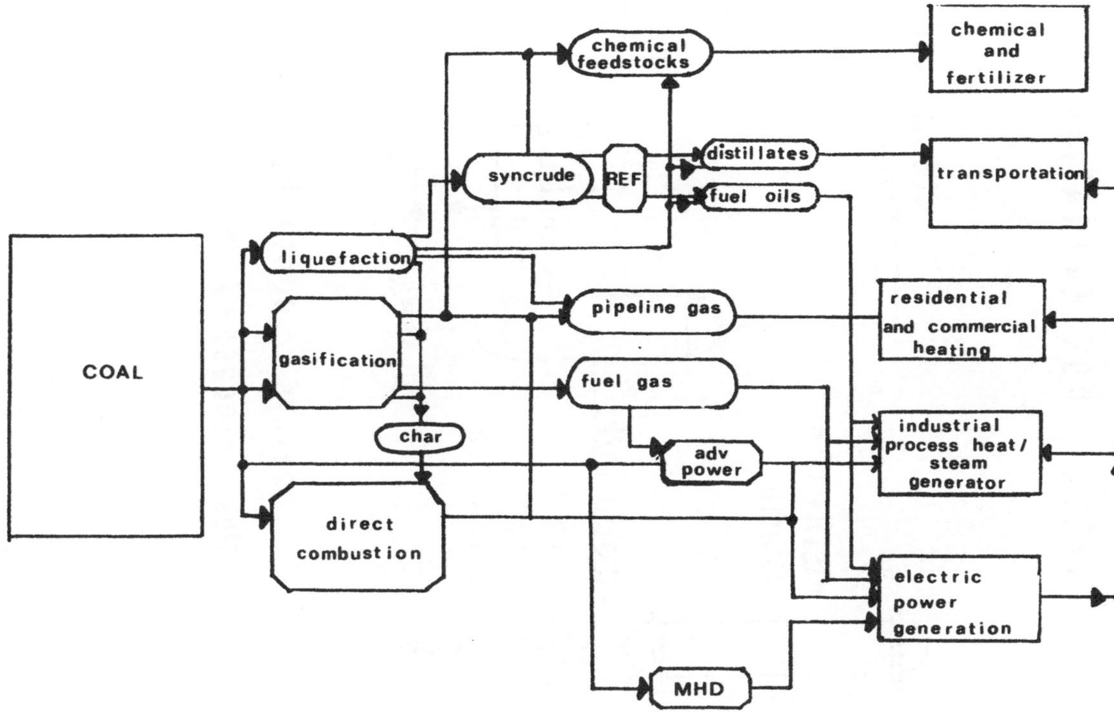

WASTE CONCENTRATION

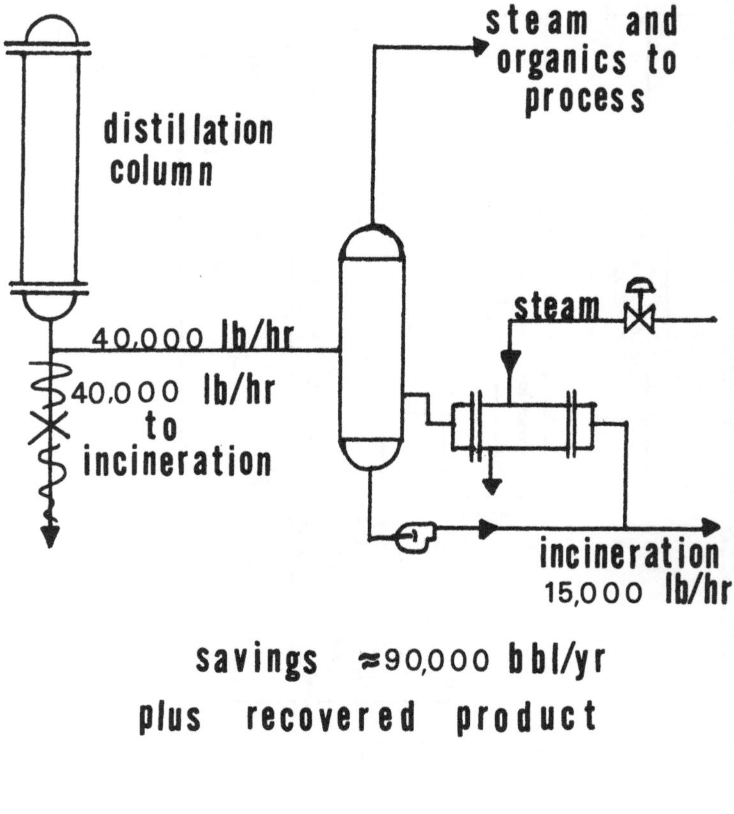

savings ≈90,000 bbl/yr
plus recovered product

figure 5

BACK PRESSURE TURBINES

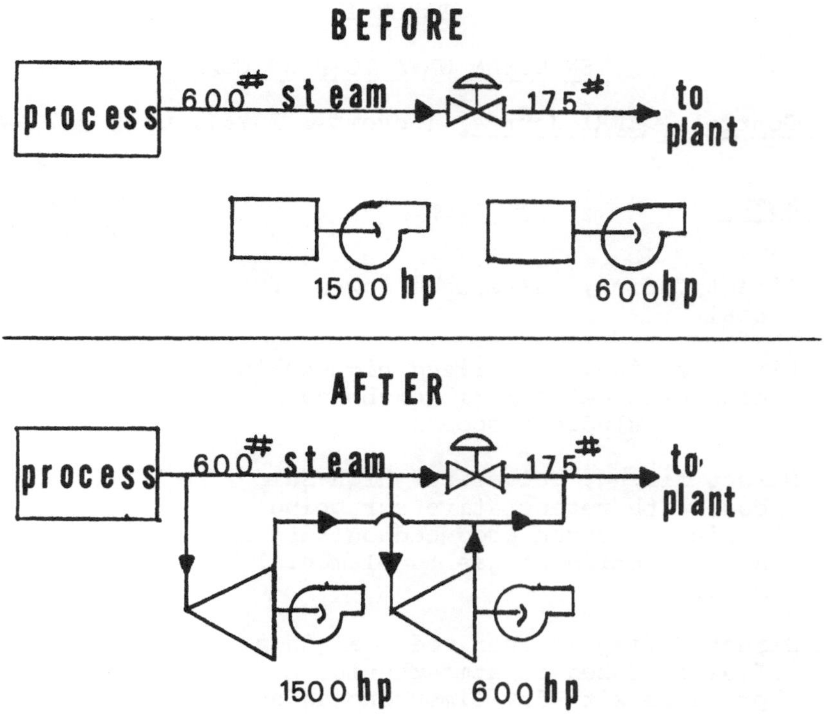

EQUIVALENT ENERGY SAVINGS 23,000 BB/YR

figure 6

TABLE I

TEN CLEAN WAYS TO BURN COAL

Source: *EPRI Journal* (November 1976), p.9.

METHOD	SULFUR REMOVED (%)
Direct firing, pulverized low-sulfur coal.	None
Direct firing, pulverized high-sulfur coal with wet alkali scrubbing and on-site sludge disposal.	80-90
Direct firing, pulverized high-sulfur coal with regenerative scrubbing (using hydrogen produced on site) and reduction of gas to elemental sulfur.	90-95
Direct firing, pulverized coal in a fluidized bed at atmospheric pressure with dry limestone added to bed.	80-90
Direct firing, pulverized solvent-refined coal.	70-90
Liquid firing, petroleum-type fuel.	90-95
Low-Btu gas firing (100-150 Btu/scf) after gasification in moving-bed, dry ash Lurgi gasifier and Selexol process for acid gas removal.	95
Meidum-Btu gas firing (300 Btu/scf) after gasification in slagging moving-bed gasifier and Selexol process for acid gas removal.	95

METHOD	SULFUR REMOVED (%)
Low-Btu gas firing (100-150 Btu/scf) after gasification in atmosphere, two-stage entrained gasifier and Stretford process for acid gas removal.	95
Medium-Btu gas firing (300 Btu/scf) after gasification in pressurized, two-stage entrained gasifier and Selexol process for acid gas removal.	95

Appendix

EXAMPLES OF ENERGY CONSERVATION BY DU PONT

Liquid-Liquid Extraction

Distillation is the most common means used by the chemical industry for separating a mixture of liquids into its components. Large amounts of energy are required, usually in the form of steam, to bring about this separation in a distillation column.

There are other chemically sound means of accomplishing this separation, but in the days of low energy cost, distillation was almost always more economically attractive. One of the alternatives, liquid-liquid extraction, is a process whereby a liquid having high affinity for one of the components in the mixture, but immiscible to the balance, is introduced into the multi-component fluid where it combines with the desired component and is separated from the mixture by decantation.

Recently, this process was employed to recover a solvent utilized in the manufacture of a synthetic fiber. Prior to this installation, several alternatives were investigated. After small-scale tests demonstrated the feasibility of the process and the acceptable quality of recovered solvent, the extraction route was selected, primarily for its lower energy costs. If a straight distillation scheme had been installed, it would have required the equivalent of 40,000 barrels of oil per year more than the extraction process.

Several other fiber plants, now in production, use straight distillation to recover the solvents from aqueous solutions. Study is now underway to determine the justification for installation of an extraction process to recover this solvent. Extraction solvents have been tested and the best one has been selected. For the one fiber plant where this extraction scheme is most likely to be installed, the associated annual savings in energy are equivalent to 100,000 barrels of oil.

Removing Water from Product

One step of a manufacturing process for a new product made it necessary to provide equipment to dry a water-saturated web of material. The original concept merely called for passing the web through a steam-heated dryer. A design study showed that significant steam savings would result if a water press were installed ahead of the dryer to "squeeze" some water from the web. Operating cost savings from reducing steam requirements in the final facility design were sufficient to justify the additional investment in the water press. Energy savings were equivalent to nearly 28,000 barrels of oil per year.

Purchased vs. Self-Produced Oxygen

A plant normally operated two oxygen production facilities. Recently, requirements were less than the minimum turndown capabilities, with the result that the excess was vented to the atmosphere. To avoid this condition, one unit was shut down, the other operated at full capacity, and the balance of the requirements were purchased. Elimination of venting saved the equivalent of 28,400 barrels of oil per year.

Waste Heat Recovery

An economizer was installed in the breeching from four boilers in addition to the economizers already existing in each boiler. The additional heat recovery saved the equivalent of 23,700 barrels of oil per year.

Optimization of Column Operation

One chemical plant revised the method of operating refining columns, to run some at full capacity and to shut down others. This saved the equivalent of 10,000 barrels of oil per year compared with normal procedure of running all units at partial capacity.

Waste Disposal

It was the practice of one textile plant to dispose of a waste finish oil in the powerhouse boilers. Since the oil was in a highly aqueous solution, this was an inefficient procedure. A chemical recovery process was developed to extract the oil and save the equivalent of 8,400 barrels of oil per year.

Maintenance Procedures

An intensified maintenance program on one plant to repair steam and compressed air leaks as soon as possible saved the equivalent of 7,800 barrels of oil per year. By the former procedure, leak repair had a lower priority on maintenance schedules.

Improved Instrumentation and Controls

Additional instrumentation and controls were installed on a chemical plant to permit operation of a direct-fired steam superheater with a minimum of excess air. This saved the equivalent of 4,000 barrels of oil per year.

Shutdown of Spare Equipment

Steam jets are used to create a vacuum at a textile plant. Instead of running spare jets to maintain reliability, an automatic start system was installed for quick start-up, permitting shutdown of the spare units when not in use and resulting in a saving equivalent of 3,000 barrels of oil per year.

Energy Savings Through Waste Incineration

A distillation column (see attached sketch) at one of the chemical plants produced 40,000 lb/hr of highly aqueous waste (95 percent water). Incineration was the only environmentally sound method for destroying this waste, but large amounts of natural gas were required to evaporate the water so that the organic material could be destroyed. Our engineers conceived a scheme whereby a concentration still was installed that, through evaporation, reduced the waste load to the incinerator from 40,000 lb/hr to 15,000 lb/hr. The remaining 25,000 lb/hr was discharged as pure water vapor off the top of the still where it was reutilized in the distillation process as a source of heat. This installation is now in operation and is saving the equivalent of approximately 90,000 barrels of oil per year.

Energy Savings Through Turbine Drives

At one plant, a process involved cooling by evaporating water at high pressure (see attached sketch). Almost all of the resulting 600 psig steam was reduced to 175 psig for process use, with only a small stream used at the higher pressure. During the design for an expansion of this process, the engineers conceived a scheme where this high pressure steam could be utilized to drive turbines that would exhaust at 175 psig. The existing generation plus that which would be brought about by the expansion was sufficient to replace motor drives on two pumps, totaling 2100 hp. This scheme was adopted and the turbine drives are now in operation, resulting in a savings equivalent of 25,000 barrels of oil per year.

Reuse of Exhaust Air

The alertness of a designer during his visit to a textile plant has paid off in energy operating cost savings. During this visit, the designer noted that some simple changes made on two existing ventialtion systems could save energy.

He observed that 50 percent of the ventilated air from a shop and substation was recirculated and the remaining 50 percent was exhausted to the atmosphere at 75°F. He reasoned that the air being exhausted to the atmosphere could be diverted during the winter months to supply 40 percent of the outside air required by an adjacent air conditioning unit serving the spinning and washdown areas. This would reduce the amount of steam required to preheat the outside air. Two new exhaust fans and some ductwork and dampers would be needed to implement his idea.

He sold the proposal by showing a 65 percent return on investment after taxes that required a modest investment of approximately $10 million. Nearly 1,400 equivalent barrels of oil will be saved as a result of this change.

Energy Savings in Distillation

In an attempt to reduce investment costs, two abandoned distillation columns from an obsolete process were reused in a new process application. It was found that in order to meet project objectives, energy requirements would be two times that required for a "neat" design for one column and five times for the other. Further investigation disclosed that by binding particular plates on both columns, they could be brought into line with flowsheet requirements. It was also determined that modifications made on the column interiors to reduce the energy use while meeting process requirements could be accomplished for $9,000. Potential annual operating cost savings of about $40,000 overwhelmingly justified this $9,000 expenditure. If maximum energy conservation is achieved through this effort, total energy reduction will exceed 4,000 equivalent barrels of oil per year.

Summary of Recommendation Subject Areas

Principles Guiding the Rational Use of Coal

II.A	Internalization of Social Costs - General Principle
II.B	Energy Pricing
II.C	Internalization of Air and Water Pollution Costs Through Governmental Control Strategies
II.D.2.1	Phasing Out of Gas Use for Steam Generation
II.D.2.2	Phasing Out Petroleum Use for Steam Generation in Large Units
II.D.2.3	Energy Conservation Technology Transfer Service

Near Term Coal Utilization

III.A.1	Base Coal Conversion Programs on Availability of Coal
III.A.2	Coal Conversion Conformance with Air Pollution Regulations
III.A.3	Exemption of Small Units from Mandatory Conversion
III.A.4	Stability of Regulatory Environment
III.A.5	Importance of Case-by-Case Analysis

Cogeneration

III.B.1 Reformulation of Department of Energy Rule; Reformulation to Encourage Cogeneration

III.B.2 State and Federal Policy Initiatives to Encourage Cogeneration

III.B.3 Base New Source Performance Standards for Cogeneration on System Output, Not Input

III.B.4 Require Actions on Cogeneration by Federal Agencies, Including Power Authorities

III.B.5 Develop New Personnel Training Programs

Research, Development, Demonstration and Commercialization

IV.1 Government has a Major Role in Promoting Research and Development

IV.2 Plants for Commercial Use are not the Responsibility of the Government

IV.A.3 Importance of Both Environmental and Economic Objectives in Selecting Technologies for Development

List of Acronyms and Abbreviations

FUEL UTILIZATION AND CONSERVATION TASK FORCE

bbls/D - Barrels per Day

DOE - Department of Energy

ERDA - Energy Research and Development Administration (now part of DOE)

GW(e) - Gigawatts (electric) = 1,000 MWe

psi - Pounds per square inch

WAES - Workshop on Alternative Energy Strategies

Glossary

FUEL UTILIZATION AND CONSERVATION TASK FORCE

anthracite coal - hard coal; coal of the highest rank.

average cost pricing - a process whereby tariffs are based on historic costs such that the most recent costs associated with production are averaged with costs based on actions taken in prior years.

bituminous coal - soft coal; a category of coal intermediate in rank between lignite and anthracite.

bottom cycle - a system (such as an electrical generator) which operates on low grade steam (i.e., low temperature); normally steam of this grade would be rejected by an electrical power station or industrial process.

closed (or indirect) cycle - a type of electrical generating system in which a heat transfer system is regenerated internally. This type of cycle has a higher thermodynamic effect than open cycles.

coal baghouse - cloth filters which physically remove particles from the effluent stream of a combustion process. Similar to a vacuum cleaner bag, only much larger.

cogeneration - the process during which energy in the form of heat is converted into electricity or mechanical energy through the use of turbines, internal combustion engines, or other heat engines, and in turn, the resulting exhaust from this conversion process is utilized as a source of energy for a manufacturing or processing operation, and for space heating. Cogeneration also may be defined by reversing the process described above (eg., bottoming cycles).

fluidized bed combustion - the process whereby air is forced upward through a bed of coal and limestone, which is fired by a gas flame causing particles in the bed to fluidize. Cooling water is fed through the system and can be used to produce steam for heating or electrical generation.

gasification - the process of converting a solid or liquid fuel into a gaseous fuel.

incremental pricing - pricing based on the most recent change in costs, or projected change in costs, as opposed to pricing based on historic (actual) costs.

lignite - brown coal; the lowest rank coal.

oil slurry - the liquid (oil) transportation medium through which coal is transported through a pipeline. The two products, oil and coal, are then separated out at the receiving end of the pipeline.

peak-load pricing - a tariff design whereby the price to consumers is made higher during time intervals in which utilities experience peak loads than during other periods of time.

resistive-space heating - a process whereby electrical energy is transformed into heat energy for space heating.

social costs/benefits - costs and/or benefits resulting from an economic activity which are borne by, or accrue to, individuals not directly involved in the economic activity.

standby charges - monthly charges paid for the right to use a certain amount of power if needed, whether or not any is actually used. A company with its own generator may find it advantageous to pay standby charges so as to be assured of electricity should its own generator fail or require maintenance.

time-of-day pricing - a tariff design whereby as the cost associated with the production of electricity or gas varies over particular time periods, the price to the consumer is made to vary in similar fashion.

turn-key construction - a construction contract which specifies that a new plant will be delivered complete, ready to go. The owner need only "turn the key."

vintaging of customers - a proceedure under which the amount paid for a particular product depends on when the contract was signed. With buildings, the concept is obvious - a building bought long ago cost much less than an identical building constructed today would. Vintaging contrasts with rolled-in, or average cost, pricing.

SECTION 4
ENERGY PRICING TASK FORCE REPORT

TASK FORCE MEMBERS:

Alan W. Beringsmith,
Co-Chairman

Grant P. Thompson,
Co-Chairman

Ernst R. Habicht. Jr.

Cliff Jensen
Paul Joskow
Philip Mause
George Petri
Larry L. Schedin
Edward V. Sherry
Richard E. Weber

ADVISOR:

Jay Kennedy

TABLE OF CONTENTS

ENERGY PRICING TASK FORCE REPORT

I. Summary of Task Force Report

The richness of the following recommendations cannot be appreciated without reading the full report. However, for the reader's convenience, the summaries that are provided here state the general structure and conclusions of the report:

Introduction

In this section, the Task Force sets out the history of its deliberations, argues that prices are highly significant in reforming the nation's energy situation, and describes the various methods that are currently in use in the United States for establishing energy prices.

Pricing and the Market for Steam Coal

The steam coal industry, despite the fact that it includes a number of very large producers and that some of these producers are controlled by petroleum companies, nevertheless contains so many producers, including numerous independents, that it is still workably competitive. Prices are determined by the operations of the market and, assuming that there are no significant changes in market structure, any future price regulation by government appears unwarranted.

Coal and Electricity

In this section, the Task Force discusses the present regulation system in this country. Task Force members agreed on the following two shortcomings of the present rate regulation of electricity:

1. The usual electric rate structure, undifferentiated as to time-of-use, fails to convey to the ultimate consumer the fact that electricity costs different

amounts to produce and distribute, depending on when it is used.

2. The price at which energy, including electricity, is sold to the ultimate consumer affects the actions taken either to increase or decrease energy consumption.

Industry Task Force members had a number of suggestions for reforming the pricing structure of electricity that were unacceptable to the environmental members, and the environmentalists had a number of suggestions that were unacceptable to industrial members. In both cases, the report sets those suggestions out in an abbreviated way. Instead of accepting either position, the Task Force agreed on the following method for setting rates:

1. A cost study would be performed for each utility to indicate system incremental costs at different times of the day and year. Essentially, the study would have two parts:

   a. The calculation of marginal running costs for different hours of the day and days of the year and for different voltage levels;

   b. The calculation of incremental capacity costs for different voltage levels at different times of the day and the year.

2. The utility's revenue requirement would be allocated among different customer classes, preferably defined by voltage level, according to the methods traditionally used by the particular utility and commission. This approach would eliminate transfers of revenue from one class of customers to another.

3. The rating periods and relative peak/off-peak rate differentials would be developed for each customer class from the results of the cost analysis and, consistent with the regulatory commission-determined revenue requirement for each customer class or voltage level, rates

would be set. The result would be that while the rates would not be set at marginal cost, the ratio between the peak and off-peak charges would be determined from that analysis.

4. Given the determination of rating periods and relative differentials, the determination of the specific rate form would be based upon the cost analysis performed in step one. That is, within the regulatory commission-approved revenue constraint for each customer class, an effort would be made to structure the charges per kilowatt hour and per kilowatt so that they reflect incremental costs. If, for example, in order to reflect such costs accurately, declining block rates were appropriate, then they would be implemented within a given customer class. Conversely, if such declining block rates obscured proper pricing signals, then such rates should not be used.

5. If the cost analysis described in step one indicates that time-of-use pricing is appropriate, then it should be implemented (over a reasonable period) for every customer for whom the benefits (calculated by estimating the consequent shift in load) exceed the additional costs of metering and administration. For all other customers, regulatory commissions may want to consider making time-of-use pricing optional, requiring customers so electing to pay metering costs. In the long run, each customer's bill will change as a function of his or her load pattern compared to the average of the class to which the customer belongs. The impact on individual customers will vary; customers whose pattern of use is close to the average will see little variation, while high on-peak and high off-peak users will see bills increasing and decreasing, respectively.

Cogeneration

The Task Force agreed that cogeneration be employed when it is economically justified. The following method of calculating the proper rates was recommended by the Task Force:

1. The rate charged for standby power should not be derived from an assumption that all cogeneration facilities may drop off the line simultaneously. The same principles used by the utility to plan for its own facilities should be applied; namely, a probabilistic analysis should be performed in which loss-of-load probabilities are calculated and decisions made which keep them at an acceptably low (but not zero) value.

2. The price paid by the utility for power fed into its grid from cogeneration facilities should reflect both capital and operating cost savings to the utility. Again, a probabilistic analysis should be performed to determine such savings.

Pricing and Social Policy

Rising energy prices will inevitably have an adverse impact on some groups of energy users. However, subsidies for energy rates, such as lifeline rates, should not be used to redress the social inequities resulting from increasing energy prices. Such subsidies have the effect of encouraging rather than discouraging energy use. A better approach to these social inequities would be through programs designed to conserve energy through efficiency improvements (e.g., through subsidies, regulations, or other devices for improving building insulation, increasing appliance efficiency standards, etc.). Negative income taxes and other such direct cash payments should also be considered in achieving a solution to the equity problem.

Issues in the Pricing of Coal Gasification

The Task Force was unable to come to any definite conclusions concerning how gas derived from coal should be priced. However, the Task Force recommends that this is an important problem to which further time and discussion should be devoted.

II. Introduction

The Energy Pricing Task Force of the National Coal Policy Project was conceived to study energy pricing policies and, where needed, to recommend changes. The Task Force has considered not only the pricing of coal, but also the pricing of electricity and synthetic natural gas derived from coal. In fact, most of our discussions concerned electricity pricing. This report contains the conclusions of the Task Force with respect to these pricing issues. The conclusions are the result of give and take between environmental and industrial representatives. While they are an improvement over existing policies, they are within the realm of political feasibility.

A. History of the Task Force's Deliberations

The pricing of energy was not a significant issue during the decades prior to 1970; indeed, the real price of most energy forms gradually declined. Although there were some conflicts and warnings of shortages to come as a result of pricing policies (particularly as to natural gas), on the whole the general decline of energy prices led few to question the way in which prices were, or ought to be, determined. But in the early 1970s, the combined effects of inflation, the reduction in easily accessible domestic supplies of energy, the quintupling of world oil prices by OPEC, and the rapid growth of energy demand led to energy price increases of a magnitude never seen before. As a result, politicians, industry leaders, and regulators suddenly found themselves having to justify energy decisions affecting pricing to angry consumers. Energy decision-making was wrenched from a time in which few persons cared about the esoteric issue of pricing into a time in which everyone cared.

As businesses and individuals reacted to higher energy prices, the imperative for efficient energy use became obvious. It was recognized that pricing is the starting point from which more efficient use of energy can be achieved, and the issue was discussed at the National Coal Policy Project's initial session at Airlie House in July 1976.

At that first session, environmental representatives argued that prices for all goods and services should be market-determined where a workable competition exists. In the case of the regulated electric utility industry, they argued that prices should be set at marginal cost. The environmentalists maintained that such marginal cost pricing should reflect the additional cost to society of producing an additional unit of energy or conversely, the savings to society resulting from not having to produce an additional unit of energy.

Industrial representatives expressed a different perception of how market prices are determined, and as a result had severe reservations about setting electric rates at marginal cost. However, they felt that there are problems with existing methods of energy pricing and that a resolution to achieve the general goal of improving the efficiency of resource allocation was worthy of further study. Thus, while there was disagreement over specifics, both sides agreed on the importance of further exploration of areas of agreement and disagreement relative to energy pricing.

The Task Force held seven meetings from February through October, 1977. This report is the product of those meetings.

B. The Significance of Prices

According to both economic theory and common observation, prices send information to buyers about the dollar cost of the resources devoted to producing a given product. Consumers who respond to those price signals (by deciding either to buy or not to buy) convey information to producers as to whether production should be increased or decreased. Thus, prices help assure that the goods that consumers demand at certain prices will continue to be produced. Prices also play a role in allocating resources efficiently. If goods are priced at a level which includes all the costs of production (including social and environmental costs), then we can expect that resources will flow to the production of these goods in an "efficient" manner. In other words, if customers receive a signal of the true cost involved in producing a good or service and are willing to pay that price, then that good or service will be produced. If, on the other hand, consumers are not willing to pay the total cost of producing a given commodity, then the commodity cannot continue to be produced at a profit by a firm, unless the production is subsidized. Such subsidies disguise the true cost of production and result in too many resources flowing to the production of this particular good, and too few flowing to the production of goods providing the subsidy.

Improper energy prices have an even more insidious effect on the nation's energy consuming or conserving investment decisions. Investment decisions are made at the margin, i.e., the separate impact that each particular alternative investment might have on the total cost of operation is assessed individually. Normally, the alternative with the greatest yield relative to its cost is chosen, all other things being equal. However, for investment decisions (especially long term ones) to result in the efficient use of society's scarce resources, those decisions must be based on accurate information. The most important piece of information relates to price. If investment decisions are based on a cost of energy that is less than society's cost of production, then these investment decisions will result in capital investments that are more energy intensive than they should be. Thus, efficient investment decisions and the overall productive capability of the economy are critically dependent on price.

Investment decisions are not confined to the industrial sector, of course. The energy efficiency of residential buildings is also adversely affected by inaccurate pricing. Just as business investment decisions may be skewed, so too are decisions as to residential appliance choice, amount of insulation, and other energy conserving options. The economic viability of these decisions is a function of the price of energy.

Prices are also important in that they encourage desirable innovation. If prices are kept artificially low through subsidy or deliberate government policy, then invention of more efficient energy using devices will slow down as innovators face a reduced profit potential for an array of items ranging from new industrial processes to solar-powered lighting devices.

Price also serves as a rationing device that balances supply and demand. When supply and demand are not in equilibrium, an unconstrained price will either rise or fall, thereby reducing or increasing demand while increasing or reducing supply until the two are in balance. Price regulation thwarts this balancing role so that in periods of shortage, such as that which the nation experienced in past winters with natural gas, there is an excess of demand over supply. The existence of such unfulfilled demand means that some sort of non-price allocation or rationing must be used to determine which consumers will obtain the supplies they desire and which consumers must be denied. This administrative or bureaucratic approach to allocation is inefficient in an economic sense, and is also frequently grossly unjust to individuals with special needs or desires.

The fact that the price mechanism is relied upon to distribute energy is value-neutral to the extent that the existing income distribution in the society is taken as a given. If prices include all costs of production, then the price system is the most efficient way of insuring that the amounts and types of goods desired by the consumer are produced.

The acceptance of the price system as an allocation or rationing device does not imply that members of the Task Force accept or reject the present distribution of income in this country. We simply urge that the issues of energy pricing and income distribution be analyzed separately.

C. Current Approaches to Determining Energy Prices

Energy prices are presently determined in a number of different ways. These differences can be explained by differences in market structure, by the nature of the commodity, by differences in government policy, and by history and tradition.

The coal sector of the economy, with its many small producers and unregulated prices, is quite competitive. The oil and natural gas sectors have large numbers of producers, concentrated refining and distribution networks, and regulated prices; yet many observers judge the market to be workably competitive. The sectors dealing with the production of and distribution of electricity and distribution of natural gas are monopolies, completely price regulated.

In the unregulated competitive sectors, such as coal, prices are determined by the market, and the result is generally consistent with economic efficiency. Thus the Task Force recognized the need to concentrate its efforts on pricing in the energy sectors related to coal where free markets do not exist.

In the case of oil and natural gas, elements of both regulation and market determination coexist. Congress is currently wrestling with the issues of oil and natural gas pricing and we have little to say on this subject.

The pricing of coal, synthetic natural gas, and electricity is the subject of this report. The bulk of the coal produced in this nation is used in the production of electricity. Assuming that coal production takes place in a workably competitive market, the price of electricity is critical to decisions related to resource allocation, investments in coal production, transportation, and conservation investments, and finally, to innovation. If the price of electricity does not include all relevant costs, excess quantities of coal will be mined and burned with resulting economic, social, and environmental impacts.

The Task Force's extensive discussions of the retail pricing of electricity included consideration of the following issues: marginal cost pricing, time-of-use pricing, embedded cost pricing, mechanisms to insure that utility revenue requirements reflect current and future costs, the internalization of social costs and benefits, utility investment on the customer side of the meter, and emission charges. The discussion contained in this report attempts to describe and analyze the strengths and weaknesses of each approach, and to recommend action where agreement was reached.

III. Pricing and the Market for Stem Coal

A. Introduction

The dramatic increase in the price of coal over the past four to five years has resulted in charges that the coal industry is not a truly competitive industry. There are suggestions that coal prices are rising solely to track oil and natural gas prices on a "what the market will bear" basis with no relationship to the cost of production.

We believe that the coal industry has been, is today, can be, and should be allowed to remain, a competitive industry without the intervention of the Federal government in the form of price controls or allocation authority. As this chapter suggests, the rise in the price of coal proceeds not from some anti-competitive forces at work, but rather from the very workings of the free market that members of this Task Force have endorsed as a general principle or basis for energy pricing.

B. Pre-1970s Coal Prices

Throughout the decades following World War II and until the late 1960s, the United States coal industry was faced with little or no growth and an excess capacity. Some of the diminution in the importance of coal was a result of increasing concern for the environment; however, a major factor in this trend was the competition from cleaner, apparently lower-priced oil and natural gas. An increasing awareness of the social costs associated with the use of coal led to decreased demand and discouraged additional investment in existing or new mines. This decline in demand made it unnecessary for the coal industry to generate funds for new, and presumably higher cost, capital investments. This situation in turn helped produce the stability of coal prices in current dollars and decreases in constant dollars. In 1969, the average price per ton of coal sold in the United States was $4.99, exactly what it had been in 1948. However, when expressed in 1948 dollars, the average price of coal in 1969 was only $3.06 per ton, nearly a 40 percent decrease. Perhaps even more significant to this price trend than the lack of need for new capital was the coal industry's ability, over a period of 20 years, to increase productivity measured in tons of output per man-year. This increased productivity offset the effects of increases in wages, capital costs, and general inflation.

C. Recent Coal Price Increases

Several things occurred which help to explain the price increases of the 1970s. Just as the costs of damage to the enviornment were internalized through restrictions on the use of coal, the passage of the Coal Mine Health and Safety Act of 1969, federal leasing policies, and Mineral Leasing Act amendments have forced internalization of the social costs associated with the health and safety of the coal miner and the disruption of the environment. It is important to note that the extra costs now paid by the users of coal are not new costs; instead, they are costs that were formerly borne by workers or persons living around the mines. The new standards have reversed the historic trend of increases in worker productivity, using current technology, thus making labor and other operating costs per ton higher. Labor costs are also likely to rise as a result of the need to attract additional workers to the industry. It appears that, at least in the short run, there may be significant shortages of labor. These higher costs, in turn, will result in higher prices for coal.

Another significant factor affecting coal prices is the cost of and the need for new capital investments in the coal mining industry. A trend of increased demand for coal began in the late 1960s. The coal industry could no longer continue to meet demand from existing capacity and was required to add new capacity. To provide the industry with a reasonable rate of return on this new, higher cost investment, prices increased. With forecasts projecting a doubling of coal demand over the next 10 to 15 years, it may be anticipated that real prices will continue to increase in order to finance the associated expansion.

D. The Pricing of Coal

For the sake of simplicity, coal prices are generally reported as average contract and average spot prices. This frequently leads to a misunderstanding of exactly what the price of coal is at a particular time. Prices actually vary widely and depend primarily on the cost of mining, but also on the quality of coal and the transportation costs. Although the location of the coal deposit and its quality and quantity will influence price by affecting the rate of return that the coal producer is able to justify, the overriding factor determining the price of coal is the cost of production. In fact, it can be said the long run coal price generally tends to be a function of the weighted average cost of new mines.

Approximately 80 percent of the coal sold in the United States is sold under contract. The terms of these contracts range from several years to periods of 20 years and longer. Most older contracts (in existence ten years or more) are so-called escalated contracts. Initial, or base, prices were established by estimates of expenditures (capital, labor, materials, etc.) required to begin mine operation. However, provision is made for the contract price to escalate with changes in the cost of labor and capital. In general, the newer the contract, the higher the price will be. This situation results from the interaction of two factors. First, there is the increased cost of capital, noted above. The investment required for opening a mine in 1977 is several times what it would have been five years ago, and these higher costs must be amortized over the life of the mine. (It should be pointed out, however, that the higher prices for coal could conceivably call forth innovative mining technologies that will drastically alter traditional cost estimates.)

The second factor making the price of newer contracts higher than the price of older contracts is that much of the coal that is cheap to mine has already been mined. Coal occurs in varying seam thicknesses, qualities, and depths. The highest quality, most easily mined coal deposits are naturally worked first. Over time, new mines are generally less productive, smaller, or of lower quality. Each of these factors affects the final price of coal.

As a result of uncertainties associated with the cost of production, the escalated contract is, with some modifications, the basis for most new coal supply contracts. The most significant modification is the "market re-opener." The re-opener provides for periodic base price renegotiations to bring the base price in line with that for newly negotiated contracts for similar quantities and qualities of coal. Re-opening takes place typically every three to five years. This clause is a response to the rapid increase in capital required to finance new mines, to replace old mines, and to expand or open new mines.

Only 20 percent of the coal produced in the United States is sold on a non-contract spot basis. The spot market is a residual market that is significantly affected by short run disequilibriums between supply and demand. As a result, the spot market price fluctuates substantially according to short-term conditions of supply and demand. Because of the lengthy lead times required to open new mines (three to five years for large surface mines and longer for deep mines) the industry's response to increases in demand is not immediate and this is reflected in the wide and sometimes rapid changes in the spot prices for coal.

Today the contract price for coal varies from under $4.00 per ton for surface mined western coal to over $30.00 per ton for deep mined eastern air pollution compliance grades. A large utility might report its average cost at $22.00 per ton, for example, but individual prices it pays will vary widely. It is important to note that average coal prices cannot be directly compared with prices for fuel oil. Competition between coal producers generally determines the prices, which are usually not uniform. On the other hand, crude oil prices, with minor variations for quality, are generally uniform in price, as determined by the market or government. Because of the nature of the coal market, an average price for coal cannot be meaningfully compared to oil or natural gas prices. In general, the coal consumer with old contracts will have an average coal acquisition cost well below those consumers with recently negotiated contracts.

E. Competitiveness of the Coal Industry and Ease of Entry

Several studies conducted by the Federal government have concluded that the United States coal industry is workably competitive.1/ A study by the Interior Department makes the following statement:

1/ See Council on Wage and Price Stability, "A Study of Coal Prices." See also, U.S. Department of Interior, The State of the U.S. Coal Industry, Information Circular 8707.

> The bituminous coal industry is highly competitive as to price, service, and quality of product and is subject to competition from other non-coal energy sources. The top 15 bituminous coal-producing companies mined about 46.6 percent of the nation's total production in 1974, and the balance of approximately 3,885 companies were responsible for the rest of the production. Thus, the U.S. bituminous coal industry is not an oligopoly.2/

A recent occurrence which has had the effect of making the coal mining industry more competitive is industry's use of captive coal production for its own needs. This has become common in the electric utility, steel, and chemical industries, which taken together consume about 70 percent of the nation's coal production. Coal mining is far less capital intensive than electric generation. The investment required to open an economically sized mine is measured in tens of millions of dollars, while the investment required for a utility or large industrial sized coal-fired boiler is measured in hundreds of millions of dollars. Despite this enormous investment, over 50 percent of the cost of operating a new coal-fired boiler can be the cost of the fuel. Thus, substantial financing incentives exist for the utility industry and other large industrial users of coal to integrate back to the coal field, if coal producers attempt to obtain rates of return greater than those available to the consumer of the coal from its own operations.

In spite of this competitiveness, the impact of increased demand, particularly in the short run, can significantly increase coal prices. This happened during late 1974 and early 1975 as users began stockpiling coal in anticipation of a strike. As a result, coal prices increased dramatically which in turn substantially increased profits. However, 1974 was an abnormal year for the coal industry as well as for many other industries. In the aftermath of the Arab Oil embargo and the United

---

2/ Ibid., U.S. Department of Interior.

Mine Workers' strike, many purchasers panicked, thus driving up process and industrial profit margins. Since 1974, conditions have tended to be more normal; in fact, despite additional requirements for capital investment, prices have decreased in many instances.

F. Conclusion

The Task Force concludes that the coal market is workably competitive and thus prices are best determined by the operation of the market and not by the government. Hence, we oppose price regulations.

IV. Coal and Electricity

A. Introduction

The relationship between the pricing of coal and the pricing of electricity to the ultimate consumer is an important one. Central station electric generation is the nation's single largest user of coal. In 1977, approximately 65 percent of total coal production was used to generate electricity. For a number of reasons, it is probably that this percentage will increase in the future. Many projections of future energy supplies in this country call for a striking increase in production of electricity as it substitutes for other fuels. In addition, national policy is now established to drive utilities away from electrical generation by natural gas and oil and toward generation by coal.

As noted in an earlier chapter, prices are important to both suppliers and consumers of goods and services, as they strongly influence investments, production, and consumption decisions. It is clear that for a large and increasing number of consumers of energy, the use of electricity is the only way that decisions regarding the consumption or conservation of coal are transmitted. If the pricing of electricity obscures the true costs of coal, then more or less coal than is economically efficient will be produced or consumed. Although the Energy Pricing Task Force recognizes that many factors in addition to the price of coal go into the pricing of electricity to its ultimate consumer, it is the effect of electricity pricing on the demand for coal that has been the focus of many of our debates.

Elsewhere in this paper, the Task Force has urged that the free market be permitted (to the greatest extent possible) to dictate the prices at which energy is sold to consumers. Unfortunately, in the case of regulated electric utilities, this free market solution is not practical. For economic, historic, and technical reasons, electric power generation, transmission, and distribution in this country is for the most part a monopoly. The supplier accepts conditions imposed by public control of rates, and classes and conditions of service. While there has been some debate as to whether the generation and transmission of electricity truly constitutes a "natural" monopoly, it is clear that the final distribution to ultimate users is a natural monopoly since it enjoys substantial economies of scale and avoids duplication of expensive facilities.

Once public regulation is substituted for the operation of the free market, prices do not work automatically to allocate goods and services in the most economically efficient manner. Instead, public regulations must try to guess what the market would do if one existed, and then mimic it (taking other social goals into account).

B. Present Rate Regulation

In most cases, regulation of both rate of return and rate design for retail sales of electricity in the United States takes place at the state level, and is affected by the actions of the Department of Energy and the Federal Energy Regulatory Commission. In some jurisdictions, retail rate regulation is conducted at the city level. In other instances, there is no direct regulation of rates, as in the case of most municipally owned systems and the PUDs in the Pacific Northwest. Finally, the Tennessee Valley Authority and a few other power authorities determine wholesale the price of electricity they generate or transmit.

The fact that regulation is conducted at different levels and is subject to different cost and load patterns is important for the Task Force's pricing recommendations. Such a regulatory framework offers opportunities to adapt generally applicable recommendations to specific cases and to use those elements that make economic sense for given utility systems. Pluralism in rate-setting authority allows experimentation and real world tests of alternative rate designs.

C. Shortcomings of Present Rate Regulation of Electricity

Designing electricity rates has always been a difficult matter. Designing them during a time of rapidly changing fuel prices, increased sensitivity toward environmental values, high capital costs, and fear about continued secure supplies of basic fuels is exceptionally difficult. Therefore, it is not surprising that the members of the Energy Pricing Task Force have been unable to agree totally as to what changes are needed in the pricing system. But there has been general agreement among all the Task Force members that the present system is no longer appropriate for today's needs.

As in all times of major change, there have been extensive (sometimes heated) debates on the subject of electricity pricing. The country is clearly in a time of experimentation, with different groups and institutions attempting to suggest or implement alternatives to the present system. The experimentation that has taken place includes the rates developed or being developed pursuant to orders of the New York Public Service Commission, the Wisconsin Public Service Commission, the California Public Utilities Commission, and other smaller scale experiments financed either by individual utilities or by the Federal government. These experiments have been dwarfed by the volume of studies that have been released, including work by the Electricity Consumers Council (ELCON), the Environmental Defense Fund, the Federal Energy Administration, and the Edison Electric Institute/Electric Power Research Institute (EEI/EPRI) Electric Utility Rate Design Study. These experiments and reports have differed, but again it is striking that all of the efforts have accepted, either explicitly or by implication, the fact that changes are required in the present rate design.

The reader of this report is referred to any of the existing studies for a full discussion of current issues and opinions on the shortcomings of the present electric rate design. But it is worth setting out in a shorthand manner the deficiencies upon which our own Task Force agrees.

First, we agree that the usual rate structure, undifferentiated as to time-of-use, fails to convey to the ultimate consumer the fact that electricity costs different amounts to produce and distribute, depending on when it is used. Because electricity cannot be stored, it must be made at the instant that it is consumed. During peak periods of use, operating costs increase as less efficient units are pressed into service and as a greater capital interest in generation, transmissions, and distribution is required. During off-peak periods, electricity is cheaper to produce because efficient plants that use relatively inexpensive fuel are available. Yet the facts on this wide variation in operating costs are rarely conveyed to the ultimate user. Instead, electricity is usually sold as if it were of uniform cost. All members of the Task Force agree that this is a major shortcoming of the present rate design, although the members differ somewhat as to the full implications of this fact to all classes of consumers.

Second, the Task Force members agree that the price at which energy, including electricity, is sold to the ultimate consumer affects the actions that he or she will take either to consume energy or to make some other use of his or her limited supply of money (to purchase insulation, for example). Therefore, prices have an important role to play in the conservation investments made by individuals and firms. When electricity is priced so that consumers are denied knowledge of the costs to society of their consumption decisions, less conservation will take place than is theoretically optimal in the economic sense. A subsidiary problem is that price signals are blurred since bills are rendered only once a month, long after individual consume/save decisions are made. Customer education and feedback devices should be encouraged.

Beyond these broad areas of agreement, environmental and industry members differ as to the extent to which they are willing to implement changes and the speed with which those changes should be undertaken. The two sides also disagree in determining the extent to which present systems or small variations to the present system are credible attempts to move forward.

D. Industry Views

Industry members of the Task Force believe that electricity rates should be based on the cost-of-service. It is the function of the rate structure to apportion system costs in a fair and reasonable manner. The test of "fair and reasonable" requires that customers pay those costs which they cause the system to incur. Thus, an important role of the rate structure is to inform the consumer of the relative costs incurred at various hours and seasons in providing his or her service as demanded. Industry believes that when the cost of providing a service varies by its time of use, the rate structure should reflect that variation in cost. If the rates deviate from cost-of-service principles, then energy and conservation choices are distorted.

Industry members believe that time-of-use rates based on embedded costs can provide the consumer with the proper price signals necessary to make rational decisions on energy consumption. They oppose time-of-use rates based on marginal computations of uncertain future costs. Time-of-use rates based on marginal costs require arbitrary adjustments to meet utility commission-determined revenue requirements; those arbitrary adjustments may well result in deviations from cost-of-service and distortion of the price signals.

Existing rates do not always indicate the variation in the utility's cost of providing service from one time period to another to the consumer. This cost may change significantly from moment to moment depending on the mix of generating units on line, the relative load on those units, the fuel utilized, and whether the utility is purchasing electricity from another supplier in order to meet load. Since it is desirable for rates to be simple to administer and understand, time periods should be grouped into relatively homogeneous blocks, thereby averaging the costs across hours within a given rating period. The most common groupings are on-peak and off-peak for time-of-day rates, and winter and summer for seasonal rates.

The economic justification for time-of-use rates is not to force customers off the system at particular times or to penalize them for consuming energy during periods of high generating costs. Rather, the objective is to insure that customers base their consumption decisions on a price that reflects the cost they force the utility system to incur. This allows the consumer to make a more rational consumption choice at a given time. Thus, generally the industry members of the Task Force favored a time-of-use rate structure based on embedded costs.

Several industry members suggested that the allowance of higher returns on equity by regulatory commissions might also make the rates more reflective of future costs as well as being valuable in signaling the cost of new plants, as long as higher equity returns are required of new plants to obtain adequate financing. Still another alternative suggested by some industry members was the inclusion of construction work in progress (CWIP) in determining the rate base in lieu of the current widespread approach which provides the utility an allowance for funds used during construction. CWIP is an accounting treatment that, by increasing the rate base and current revenue requirements, tends to more accurately reflect the replacement cost of new plants. Inclusion of CWIP and elimination of allowance for funds used during construction will increase prices and cash flow in the short run, but will not increase utility earnings. The difference is one of timing. Under the practice of recognizing CWIP in the rate base additional costs are reflected in rates when they are incurred, while the allowance for funds used during construction (AFUDC) treatment postpones rate recognition until the plant is completed, perhaps as much as a decade later. Inclusion of CWIP in the rate base will result in higher prices for rapidly growing systems and is thus one way to signal to consumers the cost of additional capital required by a utility as a result of consumers' decisions.

Finally, the industry members recognized that if the income tax savings realized by the deduction of the interest component of allowance for funds used during construction (AFUDC) was deferred, it would require prices to be more reflective of the current cost of production. Therefore, if the issue is the need to obtain prices that are more representative of current costs and if partial effects of CWIP in the rate base without AFUDC would flow to prices by deferred tax accounting, then total new construction costs could be reflected through the elimination of the AFUDC offset.

E. Environmental Views

The environmentalists believe that the proper approach to electricity pricing is to price all units of electricity sold during a particular time period at the marginal cost of producing electricity during that time period, taking into account a component for future capital required to meet rising demand. Under this approach to pricing, the ultimate consumer would feel the full impact of a decision to consume in the sense that the additional cost to society of producing the marginal kilowatt-hour of electricity would be reflected in the price. Conversely, the full benefits to society of foregoing consumption would be reflected in the savings realized by the consumer. The use of a marginally determined rate would carry with it all the benefits of economically (market) determined rates noted earlier in this report. Those benefits include efficiency of allocation,avoidance of bureaucratic rationing devices, encouragement of innovation, and leveling of costs and benefits for making non-energy consuming investments of operation changes.

As a second-best solution to the problem of inducing customers to make economically desirable investments in conservation, the environmentalists suggested that utilities should be required to make conservation investments on the customer's side of the meter and include those investments in the rate base. Since all customers would derive benefits from those saving investments (so long as they were less capital intensive per kilowatt-hour conserved than the building of a generating plant to produce the kilowatt-hour would be) all customers in the class should pay for the conservation investment.

F. Methodologies for Determining Rate Structure

It may be useful to explain briefly and in very general terms the difference between marginal cost pricing and traditional rate structure methodology. Marginal cost pricing contains both short run and long run elements. Under marginal cost pricing a charge for kilowatt-hour consumption at different times of the day and the year is derived from the marginal operating cost (primarily the fuel cost) at different times of the day and the year. The marginal cost is the cost (or savings) associated with somewhat more (or less) use of electricity at any given time, assuming no additions to the existing plant. Generally, this cost is derived by calculating the cost of placing somewhat more (or less) demand upon the generating unit that has the highest running cost of all the units on line at any given time. This marginal cost (often referred to as the system lambda) is almost universally higher than the average operating cost because an economic consideration results in the dispatch of electrical power from generating facilities in the order of increasing operating cost.

Long run marginal cost includes, in addition to the marginal operating cost, a charge for the need for new capacity (not only generating capacity, but also transmission and distribution capacity) created by the consumption of electricity at various times of the day and year. This charge (sometimes referred to as a "capacity" charge) varies during different times depending upon whether capacity must be added to accommodate growth in demand at those times. The amount of the charge is calculated by reference to the increase in costs (or savings) due to an increase (or decrease) in consumption at those times of the day and year in which such increases (or decreases) in consumption would have the actual effect of increasing (or reducing) the need for the utility to make capital investments in generating, transmission, and distribution capacities.

This calculation can be made in a number of ways. However, perhaps the simplest method is to examine the utility's capacity expansion plan, and to determine how that plan would be changed in response to various changes in load. The savings (or costs) associated with the changes in the expansion plan, which would be made due to increases (or reductions) in load at different times, are then used to calculate marginal capacity charges for different times of the day and year. When added to marginal running costs, these charges result in a set of prices for electricity that reflect the marginal costs associated with the use of electricity at different times and voltage levels. Under marginal cost pricing, the rate paid by consumers parallels very closely the costs that additional consumption imposes upon the electric utility. No reference is made to historic or embedded costs. The value of such an approach is that it helps insure that new capacity is added (and that fuel is burned) only when consumers are willing to pay the actual costs of those expenses incurred by the utility.

Traditional rate-making methodology relies on an allocation of the embedded costs of the utility. Although there is a wide variety of methodologies for allocating these costs among different customers and different usages, it can generally be stated that traditional methodology attempts to allocate these costs among customers so as to produce revenue equal to the utilities' revenue requirement. Difficulties with respect to the allocation of "joint costs," occur under the traditional method. A common means of allocating those joint costs is allocation on a peak responsibility basis. In other words, the joint costs of facilities used to serve a large number of users and usages at different times of the day and year are allocated to customers on the basis of their contribution to system peak. Other systems exist for allocating these joint costs and, depending upon which system of allocation is selected, rates for different customers may vary significantly.

Proponents of pricing based on marginal cost argue that this problem of allocating "joint costs" does not occur to the same extent under marginal cost pricing due to the fact that in theory the marginal cost principle assigns the cost of new capacity to those times of the day and year in which growth in demand would necessitate the addition of such new capacity. However, in practice, since marginal prices are "averaged" on a time period and not instantaneously determined, the problem of accurately allocating joint costs exists under this principle as well.

G. Advantages and Disadvantages of Methodologies

It was the perception of the members of the Task Force that the result of charging marginal cost for all units of electricity consumed at all times would generally be to produce total revenue in excess of the revenue required to provide a fair return on the electric utilities' investment as currently calculated by State regulatory bodies. No systematic analysis was done to insure that this was universally the case in the United States, but it was the experience of a number of the Task Force members that in most states in which the problem was analyzed it was clear that the charging would be unlikely to pro-
all units of consumption would be unlikely to produce revenue equal to that authorized by the regulatory commission. Resulting revenue may be either too high or too low. However, it is currently expected that for at least two reasons marginal cost pricing would produce excess revenue. First, the marginal running cost (often represented by the system lambda of the electric utility) is almost inevitably higher than the average running cost because of economy dispatch; second, marginal capacity costs (the costs of expanding the system) are higher than embedded costs due to inflation and the enormous increases recently experienced in the costs of new electric generating capacity.

As a result, adjustments must be made to marginal cost pricing if an electric utility is not to receive excessive revenue given such electric rate restructuring. A number of adjustments are possible: a tax could be imposed on producers of electricity; some form of the inverse elasticity approach could be applied; or some other measure could be taken to insure that the charging of marginal costs for all units consumed did not result in excess revenue to the electric utility. The presence of the excess revenue generated by the marginal cost pricing is a sensitive issue and might, in the view of some of the Task Force members, create the potential for political abuse in the reallocation of the excess revenue.

Marginal cost analysis enables the regulator to determine the propriety (or impropriety) of adopting time-of-use pricing and other rate designs (for example, interruptible service) that reflect different costs and charges at different times of the day and year. (Of course, time-of-use pricing has been adopted by some systems without a formal marginal cost study.)

Time-of-use pricing is not a universally desirable reform. For reasons of weather, seasonal variations in the availability of hydroelectric power and system design, time-of-use pricing may be inappropriate and, in fact, perverse when applied to some electric utility systems. Rate restructuring must be conducted on an individual utility system basis. Time-of-use pricing should only be applied when its application will result in the potential for economic benefits to the system if consumers respond to the peak/off-peak price signals. The benefits or costs to the system are properly calculated by marginal analysis. This is because only through marginal analysis (that is, only through the analysis of the increased costs associated with increases in use at different times of the day and the year) can it be determined whether and to what extent a system will experience cost savings (or increases) due to a decrease (or increase) in consumption at different times.

Industrial members of the Task Force agree on the necessity for marginal analysis as the basis for investment decisions. However, this does not diminish their concern over the use of prices based on marginal cost. As mentioned above, if all rates are set equal to marginal cost, actual revenue will only by accident equal that authorized by the regulatory commission and, while there may be a place for marginal analysis in electric utility rate making, pure marginal cost pricing is impractical due to the revenue constraint placed on the industry by its regulatory commissions.

Members of the Task Force were unable to reach agreement concerning resolution of the excess revenue problem. In addition, industry argued that by the time adjustments are made to a marginal cost determined price so that the actual revenue obtained is in compliance with that which is authorized, the consumer price signal, particularly his total bill, will be virtually unchanged. As a result, industry members questioned the need to employ what they perceive as an uncertain and complex costing methodology to yield a set of prices similar to that which exists under the current rate making approach.

Environmental members argued that if marginal cost pricing or at least some analysis of the marginal costs associated with the use of electricity at different times is not used, it is difficult, if not impossible, to determine whether and to what extent time-of-use pricing and interruptible service are appropriate.

H. Recommendations for Electric Utility Rate Design*

The resulting recommendations attempt to capture the usefulness of marginal analysis in fashioning peak/off-peak rates while avoiding the excess revenue problem. Basically, this recommendation describes a system by which a rate structure for a regulated utility can be derived. It is assumed that the revenue requirement has already been calculated and that the only issue to be determined is the allocation of that revenue requirement to customer, demand, and energy charges among customer classes, and rating periods, i.e., times of the day and year. The methodology would include the following steps:

1. A cost study would be performed for each utility to indicate system incremental costs at different times of the day and year. Essentially the study would have two parts:

    a. The calculation of marginal operating costs for different hours of the day and days of the year and for different voltage levels;

    b. The calculation of incremental capacity costs for different voltage levels at different times of the day and the year. (In practice these costs will be averaged somewhat, rather than determined instant by instant.)

---

* Paul Craig feels that these recommendations are acceptable only in view of the longer term energy pricing policy recommendations as expressed in the Conservation and Fuel Utilization Task Force Report. (See Section II.B.)

2. The utility's revenue requirement would be allocated among different customer classes, preferably defined by voltage level, according to the methods traditionally used by the particular utility and commission. This approach would eliminate transfers of revenue from one class of customers to another.

3. The rating periods and relative peak/off-peak rate differentials would be developed for each customer class from the results of the cost analysis and, consistent with the revenue requirement for each customer class or voltage level, rates would be set. The result would be that while the rates would not be set at incremental cost, the ratio between the peak and off-peak charges would be determined from that analysis.

4. Given the determination of rating periods and relative differentials, the determination of the specific rate form would be based upon the cost analysis performed in step one. That is, within the regulatory commission approved revenue constraint for each customer class, an effort would be made to structure the charges per kilowatt-hour and per kilowatt so that charges for each customer reflect incremental costs as closely as possible. If, for example, in order to reflect such costs accurately, declining block rates were appropriate, they would be implemented within a customer class. Conversely, if such declining block rates were to obscure proper pricing signals, then such rates should not be used.

5. If the cost analysis called for in step one indicates that time-of-use pricing is appropriate, then it should be implemented (over a reasonable period) for every customer such that the benefits (calculated by estimating the consequent shift in load) exceed the additional costs of metering and administration. For all other customers, regulatory commissions may want to consider making time-of-use pricing optional, requiring

customers electing this method to pay metering costs.

This recommendation is similar to the approach taken by the New York Public Service Commission in one of the first cases in which marginal cost analysis, in conjunction with an embedded cost revenue requirement, was used to derive a rate structure for a public utility.

In the long run, each customer's bill will change as a function of his or her load pattern compared to the average of the class to which the customer belongs. To the extent that a similarity in load patterns within customer classes exists, bills would not vary greatly from present practices. Customers whose patterns of use are close to the average will see little variation, while high on-peak and high off-peak users will see bills increasing and decreasing, respectively. Bill changes that do occur are likely to arise more from the time-of-use element than from the marginal cost analysis as we have described it. No transfer of income among classes of customers would occur through this method. Moreover, time-of-use pricing will be applied where it is economically justified and will not be applied where inappropriate. The method provides a benchmark for peak/off-peak differentials and for appropriate rates for interruptible service. Finally, it gives both regulators and utilities the opportunity to consider the actual costs associated with system expansion in making decisions among investments in conventional generation, load management, or other actions that may be appropriate in light of these costs.

While environmental members of the Task Force believe that the steps as outlined above may not be sufficient to accomplish all their goals relative to the pricing of electricity, all parties agree that the implementation of the program described in these recommendations is both a desirable and an important step forward.

V. Cogeneration

A considerable portion of all the energy used in the United States goes to satisfy the demand for heat at low and moderate temperatures (from 70°F to 500°F). Space heating and most industrial uses of process heat and steam fall in this category.

The lower the temperature at which heat is needed, the less is the useful work ("available energy" in the language of thermodynamics) that needs to be expended to provide it. Similarly, the higher the temperature of a source of heat, the greater is the available energy with which useful work can be done.

Current practice in this country dictates the use of readily available forms of energy for supplying low and moderate temperature heat and steam. "Package boilers" using natural gas or oil are widespread in industry, and space heating equipment in all sectors typically uses natural gas, oil, or electrical resistance heaters.

Such direct use of premium forms of energy is wasteful from the standpoint of energy conservation since in most cases the same quantity and temperature of delivered heat could have been provided with half or less than half of the available energy. Depending upon the circumstances, such direct use may or may not be wasteful in each instance from the standpoint of total use of resources (capital, labor, fuel, etc.).

More efficient use of available energy typically involves greater capital investment in equipment at the point of end use, for example, on-site electric generators and heat pumps. It usually results in a lower operating cost to the user, and with proper planning, lower system capital investment and operating costs for the utility.

In the case of the cogeneration of electricity with the production of process heat and steam, which is one way of using a higher fraction of the available energy in the fuels that are burned, we are concerned that many economically-justified cogeneration investments will not be made. This is because prices charged by utilities for standby power and prices paid by utilities for purchased power may not now reflect the true costs and savings to the utility and to the nation arising from such transactions.

The Task Force therefore recommends that each utility and regulatory commission assess the potential for cogeneration within its system and institute rate schedules to reflect the utility's true costs and savings. In so doing, the following points should be kept in mind:

> 1. The rate charged for standby power should be based on the same principles used by the utility to plan for its own facilities; namely, a probabilistic analysis should be performed in which loss-of-load probabilities are calculated and rates are based on that analysis. Rates should not be derived from an assumption that all cogeneration facilities may drop off line simultaneously.
>
> 2. The price paid by the utility for power fed into its grid from cogeneration facilities should reflect both capital and operating cost savings to the utility as determined by a probabilistic analysis.

The application of these principles will typically result in charges for standby power and payments for purchased power that vary with the time of day and season of the year.

It should also be recognized that these principles are more applicable to systems with a substantial number of cogeneration facilities. During the transition period from the present state of few cogenerators to a situation where their existence is more common, these general principles will still be applicable but their significance will not be as great.

The rate schedules for the sale of standby and supplemental service to customers with on-site generation, and the price at which the purchase of any excess electricity is made by utilities should contain the following elements:

> 1. The determination of rates to customers with on-site generation facilities for supplemental supplies of electricity should, along with rates to other electric customers, be consistent with time-of-use pricing principles. Each customer should

face a rate schedule that provides for higher rates for power purchased from the utility during periods of system peak demand. The fact that a given customer engages in on-site generation has relevance to the system only if his standby or supplemental consumption does not coincide with the system peak, and at any time he should face the same incremental rates as any other customer.

2. Rates paid by utilities purchasing excess industrial generated power should also be determined as a function of the time, relative to system demands and resources, during which that power is available. Ideally, utilities should be willing to pay a price for excess industrial generated power sold to the grid, that is equivalent to their savings in capital, fuel, and other costs.*

---

* However, such a rule must recognize that, in some instances, daily load curves and a large unit size may require the utility to bring on units with a system lambda higher than the price an industrial generator is asking for its power. In such a situation, it is entirely possible that the excess industrial power cannot be economically integrated into the system, as its total cost to the utility is actually higher than the price being asked. Specifically, because of the design of large baseload units it may not be possible to remove a unit with higher than the system lambda and still get it back in service to meet the next day's loads. If such a unit were taken out of service due to displacement caused by industrial excess energy available to the system at a lower price, then these savings must be assessed against the higher costs which the system would incur by not having baseload units available during periods of great demand. Therefore, we wish to caution against simplistic general rules under which such purchases must always be made.

In addition, because the need for backup capacity requirements occurs at random at both peak and off-peak times, there is an argument for a fixed standby charge for customer-owned generation. A fundamental principle of rate design should be that one group of customers is not subsidized by another. Hence, to the extent that fixed costs exist, a fixed charge is desirable.

If these steps are taken, and if other non-economic institutional barriers not addressed by the Pricing Task Force are identified and removed, we have more confidence that economically-justified investments in cogeneration will be made.

VI. Pricing

A. Introduction

The long history of government intervention in energy markets and the present national dependence on foreign energy sources have so intertwined energy pricing with the issue of social equity that it is incumbent upon the Task Force to comment on this interrelationship. Since the main thrust of this report has been towards economic efficiency, it seems desirable to clarify the distinction between economic efficiency and social equity at the outset.

B. The Issue

As energy prices rise, certain groups of Americans will inevitably be affected adversely. Concern over the impact of rising energy prices on low income groups has lead to many suggestions for relief.

The National Coal Policy Project members do not believe that it is appropriate to use energy pricing as a means to address equity issues. This does not mean that Project members are indifferent to these issues. However, such pricing mechanisms as lifeline rates can have adverse effects, especially through the stimulation of energy use which would not otherwise occur.

A better approach is to provide incentives for capital investments to reduce demand for energy. There is now abundant information, analysis, and experience to show that energy conservation measures applied to existing residential and commercial buildings can often cut energy use substantially and can be cost effective.

The Task Force concludes that approaches to equity issues based on investment in conservation are to be preferred to the subsidization of energy prices. Negative income taxes and other such direct cash payments should be used in achieving a solution to the equity problem.

C. Recommendation

Subsidies of energy rates should not be used to redress social inequities resulting from increasing energy prices. Such sibsidies have the effect of encouraging rather than discouraging energy use. A better approach to redressing such social inequities is through programs designed to conserve energy through efficiency, (e.g., through subsidies, regulations, or other devices for improving building insulation retrofit, increasing appliance efficiency standards, etc.). In designing such programs, care must be taken to assure that the intended benefits accrue to the target groups (e.g., low income renters).

VII. Issues in the Pricing of Coal Gasification

A. Introduction

In recent years, with increases in the price of competitive forms of energy, the nation has experienced expanded interest in producing synthetic natural gas (SNG) from coal. A technology has long been available but the cost of conversion has been prohibitively expensive. At present there are essentially three commercial sources of synthetic natural gas: crude oil, from which naptha, propane, and butane are derived, natural gas liquid, and coal. Virtually no SNG is produced commercially from coal as yet, although one such operation is presently under construction. About half of the commercial plants making SNG from other sources are designated by their owners as seasonal operations, used during peak send-out periods. Some observers believe the time is nearing when low Btu synthetic natural gas from coal can, for some applications, be cost competitive with other forms of energy. However, the ultimate competitiveness of SNG will be determined by the pricing approach used. Thus the Task Force determined it would be helpful to attempt to identify and analyze the major issue areas.

B. The Issues

Fundamental to a discussion of the potential for SNG development in the United States is a consideration of the way in which it should be priced. Two pricing approaches are rolled-in and incremental. A third approach, much less politically likely at this time, is marginal pricing. In the first approach, the price of the higher cost gas is averaged with the cost of existing gas. This contrasts with the incremental approach, which attempts to assign the cost of the higher cost gas directly to the user of that gas, and with the marginal approach which charges a price based on the highest cost sources for all gas consumed. The rolled-in approach is currently used by state and federal regulatory authorities, although it appears that Congress may pass a version of incremental pricing by assigning the new higher cost gas to low priority (industrial) users. The utility industry supports a rolled-in pricing approach for pipeline quality gas (high Btu) as necessary to assure a market price for SNG that is below the cost of alternative fuels. If a market cannot be assured, then it will be impossible to obtain financing for these facilities which the industry argues are required in order to gain experience with the technology. Environmentalists respond that such plants should not be constructed until there is a market for this high priced gas. However, both industry and environmentalists recognize that the issues surrounding the pricing of SNG are extremely complex and do not lend themselves to resolution under either a rolled-in or an incremental pricing approach.

C. Analysis and Recommendations

The Task Force feels that gas from these facilities should normally be priced at the full cost of production. It should be the goal of regulation to assure that prices to consumers reflect the cost of providing the gas. The Task Force recognizes that there are many unresolved issues remaining on the question of SNG pricing, and recommends that they be included in future coal policy studies.

List of Acronyms and Abbreviations

PRICING TASK FORCE

CWIP - Construction Work in Progress

EEI/EPRI - Edison Electric Institute/ Electric Power Research Institute

ELCON - Electricity Consumers Resource Council

SNG - Synthetic Natural Gas

Glossary

PRICING TASK FORCE

block rates - a tariff design whereby a price to the consumer varies with increases in consumption during a stipulated time interval.

capacity charge - a tariff design whereby a part of the price paid by the consumer is determined by the consumer's maximum instantaneous demand during a particular time interval.

cogeneration - the process during which energy in the form of heat is converted into electricity or mechanical energy through the use of turbines, internal combustion engines, or other heat engines, and in turn, the resulting exhaust from this conversion process is utilized as a source of energy for a manufacturing or processing operation, and for space heating. Cogeneration also may be defined by reversing the process described above (eg., bottoming cycles).

constant dollars - the nominal dollar value adjusted for inflation.

current dollars - the nominal dollar value.

fuel charge - an adjustment to the customer bill to compensate for wide variations in fuel costs to the electric or gas utility.

gasification - the process of converting a solid or liquid fuel into a gaseous fuel.

heat pumps - a device used for extracting available heat from the surrounding (cold) air and transferring it to indoor space.

historic costs - the actual cost in current dollars recorded at the time of purchase (accounting cost).

incremental pricing - pricing based on the most recent change in costs, or projected change in costs, as opposed to pricing based on historic (average) costs. The incremental approach attempts to assign the cost of the higher cost gas directly to the user of that energy.

internalization of social costs/benefits - the process whereby the social cost/benefit is transferred from those not engaged in the economic activity to those so engaged.

interruptible service - an electric or gas service whereby the consumer agrees in advance to the supplier's discontinuance of service during short periodic intervals.

joint costs - costs for the production of two or more commodities having the nature that the production of one automatically entails the production of the others (eg., beef and hides industries).

lifeline rates - an electric (or gas) tariff designed to have a low cost to consumers who can (afford to) only use minimum amounts.

marginal costs - the change in total costs which would result if production had changed by one unit during a particular time interval (assuming no additions to the existing plant in the case of electricity). Generally determined by calculating the cost of placing somewhat more (or less) demand upon the generating unit that has the highest running cost of all the units on line at any given time.

monopoly - a situation where a firm is the sole producer of some commodity for which there are no close substitutes, and if it faces no imminent threat of competitors.

oligopoly - an industry with a small number of large firms producing the bulk of its output.

peak-load pricing - a tariff design whereby the price to consumers is made higher during time intervals in which utilities experience highest demand than during other periods of time.

process use - a reference to the use of electricity and/or gas as a direct input to the production process (eg., as opposed to use for lighting or air conditioning).

resistance heater - a device whereby electrical energy is transformed into heat energy for space or water heating.

revenue requirement - refers to the net revenue deemed necessary by utility regulators for the payment of a return to investors of capital funds in a particular utility.

rolled-in pricing - a process whereby tariffs are based on historic costs such that the most recent costs associated with production are averaged with costs based on actions taken in prior years. Also known as average cost pricing.

social costs/benefits - costs and/or benefits resulting from an economic activity which are borne by, or accrue to, individuals not directly involved in the economic activity.

system lambda - for electrical utilities, the operating costs at any instant associated with a change in output of one kwh.

time-of-use pricing - a tariff design whereby as the cost associated with the production of electricity or gas varies over particular time periods, the price to the consumer is made to vary in similar fashion.

wing dikes - dikes that extend perpendicular from the banks of a river toward the center. They are constructed in order to maintain the center depth of the channel by trapping sediment or silt in the river.

SECTION 5
AD HOC EMISSION CHARGE TASK FORCE REPORT

TASK FORCE MEMBERS:

Jackson B. Browning
Gerald L. Decker
Laurence I. Moss
Larry L. Schedin
Bruce J. Terris

AD HOC EMISSIONS TASK FORCE

I. Memorandum: Background

A. Purpose

This memorandum will discuss the major considerations and the history of emission charges and abatement incentives proposed by task force members of the National Coal Policy Project. These proposals were initially debated by members of the Pricing Task Force and later by an Ad Hoc Task Force comprised of members of both the Pricing and Air Pollution Task Forces. These discussions led to the current proposal entitled "Recommendation of the Emission Charge Ad Hoc Task Force--An Alternative to Individual Source Emission Limitations in State Implementation Plans."

B. Initial Position of Environmentalists

Environmentalists introduced the concept of a uniform national emission charge to the Pricing Task Force. The concept was aimed at controlling certain discharges (particulates, sulfur oxides, and nitrogen oxides) from coal burning plants, and was intended to supplement the existing regulatory limit system. The proposal was based on environmentalists' concern for: (1) total loading of pollutants in the atmosphere; (2) degradation of air quality in clean air regions; (3) delays in attaining national air quality goals due to the administrative complexity of the existing system; and (4) misallocation of resources by substituting regulatory judgment for managerial judgment in determing appropriate emission levels at individual plants.

The concept was introduced to the Pricing Task Force to fill the requirement for a procedure to internalize social costs associated with the emission of pollutants to the atmosphere. Internalization of appropriate social costs will allow an emission charge to be set that will encourage cost efficient abatement. The emission charge level in the environmentalists' initial proposal was the marginal cost of abatement for an average plant at a point where the marginal cost of abatement equals the marginal benefits of reduced emissions. This determination was to be arrived at through a combination of analytical and political procedures as is now the case in the enactment of laws, regulations, and standards. These procedures consider health, welfare, and various other subjective judgments on the social costs of pollution along with information on technology and abatement costs. The setting of air quality goals (NAAQS, PSD, etc.) implicitly quantifies the socially determined external cost. The concept would also result in a residual charge even after cost efficient abatement has been achieved. According to the environmentalists, this residual charge is the social cost which users should bear when emitting pollutants into the environment.

C. Emission Charges Are Not New

The concept of a uniform national emission charge is not new. It was introduced by the Coalition to Tax Pollution, a Washington-based environmental organization, in 1971.1/ The proposal was introduced as legislation in 1972. The emission charge then proposed for sulfur ranged from 25¢ to 50¢ per pound. The merits of effluent charges have been espoused by economists for many years. A bibliography of the more significant references on this issue is included with this memorandum. This is not an exhaustive list because of the extensive literature on pollution damages and their internalization. The most extensive reference is a May, 1977 report entitled "Pollution Taxes, Effluent Charges, and

---

1/ Council on Law-Related Studies, Effluent Charges on Air and Water Pollution: A Conference Report, Reporter: Edward I. Selig, Report of Conference held on October 15 and 16, 1971 (Washington, D.C.: Environmental Law Institute, 1973).

Other Alternatives for Pollution Control" for the Committee on Environment and Public Works, U.S. Senate.[2] It is a compilation of 61 selected readings along with an overview of the subject. A wide range of terms is used to describe various forms of emission charges such as pollution taxes, pollution disincentives, effluent fees, etc. These 61 readings contain many pages of speculation by economists on how an emission charge system should work, but none report on actual trials of air emission charges as a means of encouraging abatement. On this lack of actual experience, Russell E. Train gives some good advice in his article, "Emission Charges and Other Economic Approaches to Environmental Programs."[3]

> "There is no strong need to conduct general theoretical research on economical approaches, because a great deal of work has been done and continues to be done by academics. Nor do we need studies of what problems economic approaches can be applied to (in the sense of taking a solution and looking for a question it answers). What is needed is a convincing explanation of how specific detailed economic approaches can solve specific environmental problems more effectively or at lower cost than regulatory alternatives." ....
>
> "I doubt that we have the knowledge or the experience to justify jumping into a full system of emissions charges designed as a substitute for regulations. Only by demonstrating these approaches successfully in one or a few applications to specific problems (perhaps even on a regional basis if an equitable means for doing so can be devised) will the resistance to such alternatives be overcome so that they might be considered for more widespread use."

2/ U.S., Congress, Environment and Natural Resources Policy Division of the Congressional Research Service, Pollution Taxes, Effluent Charges, and Other Alternatives for Pollution Control, prepared for the Committee on Environment and Public Works (Washington, D.C.: Government Printing Office, 1977).

3/ Train, Russell E., Emissions Charges and Other Economic Approaches to Environment Programs, Tax Notes, Volume 5, (Feb. 28, 1977) pp 92-93.

Water discharge fees are similar in concept to air emission charges. However, airsheds are not easily defined like watersheds, and polluted air cannot be conveniently gathered for treatment at a central location as can be done with water discharges. Also, long-distance drift of pollutants can cause harmful fall-out effects in remote areas even though the local ambient air quality is acceptable. Emission charges currently being used in several states are designed to recover costs of administering the regulatory limit system. Although the concept of an emission charge is not new, our investigation of air emission charges to encourage cost efficient abatement indicates that there is no current application with this intent in the United States.

D. Initial Position of the Industrialists

Industrialists agreed to explore further the emission charges if three conditions were met: (1) emission charges would replace part of the regulatory limit system, rather than merely supplement all of the existing regulatory limit system; (2) the residual charge would be eliminated;* and (3) regional differences in fuel quality and abatement costs would not cause undue hardship on industries in any one region, or result in transfer payments from customers of industry in one region to customers in another region. The environmentalists agreed that regional differences presented a problem to be evaluated. They emphasized that the entire system should be designed to motivate substantial progress toward achieving ambient air quality goals. These several agreements led to the first compromise proposal for a uniform national emission charge in which emission fees would be utilized in lieu of emission standards for individual plants. The funds collected would be accumulated separately for each major industry (e.g., electric generation, steel, etc.). Incentives to meet air quality goals would occur by refunding annually the collected revenue within each fund to members of each industry group in proportion to each member's fraction of the national output of that particular production (kwh of electricity, tons of steel, etc.). The charge would be adjusted periodically depending on progress in meeting air quality goals.

---

*A typical company that had achieved cost efficient abatement on its total operation within a given region would incur refunds equal to its charges thereby providing an incentive to abate up to the cost-effective level.

Regional surcharges could reflect regional differences in fuel quality and abatement costs.

E. Reaction by Industry Groups

There were considerable negative reactions to the national emission charge and abatement incentive proposal by various industry groups. Here is a sample of some negative reactions of the utility industry:

1. Doubt that the existing regulatory limit system which is so entrenched could really be replaced; a charge system would end up as an added burden.

2. Concern that the charge would be used as a tax and increased arbitrarily to raise general tax revenues.

3. Concern that the charge would be increased arbitrarily to prevent installation of new facilities.

4. Fear that regulators would look upon emission charges as non-compliance penalties and disallow such expenses for ratemaking purposes because of poor management.

5. Concern that the public would interpret emission charges as a license to pollute.

6. Doubt that an emission charge would be meaningful unless applied to small sources and sources other than coal.

7. Fear that costs derived from an average plant would cause inequities.

On the other hand, it is also interesting to note that some industry sectors have favored certain forms of an emission charge. This is indicated in the summary of a June 24, 1977 seminar entitled, "Pollution Charges and Public Policy in California"4/ which states that both the California Association of

---

4/ Public Interest Economics, Pollution Charges and Public Policy in California: A Seminar Summary, Summary of Seminar held on June 24, 1977 in San Francisco, CA, Public Interest Economics (San Francisco, CA, 1977).

Manufacturers and the National Association of Manufacturers support some form of pollution charges because of severe problems for locating new industry in non-attainment areas. Non-attainment provisions severely restrict opportunities for the creation of new industry in certain areas even though proposed new facilities have very low emissions. Those companies which can modify existing plants have a distinct advantage over those companies with no existing facilities in the area. Offset regulations and state attainment plans are quite ambiguous, and thereby provide a highly uncertain future for desirable industry expansion in non-attainment areas.

These reactions suggested that agreement might more easily be reached on a more limited proposal focusing on non-attainment areas.

F. Clean Air Act Amendments

The Clean Air Act Amendments of 1977 were passed during the negotiations of the Task Force. Rather than negate the negotiating effort these amendments offered new opportunities regarding appropriate applications of emission charges and abatement incentives. In addition, Section 405 of the amendments encourages further investigation of economic measures for controlling air pollution. Section 405 states:

> "(a) The Administrator, in conjunction with the Council of Economic Advisors...shall undertake a study and assessment of economic measures for the control of air pollution which could--
>
> (1) strengthen the effectiveness of existing methods of controlling air pollution,
>
> (2) provide incentives to abate air pollution to a greater degree than is required by existing provisions of the Clean Air Act (and regulations thereunder), and
>
> (3) serve as the primary incentive for controlling air pollution problems not addressed by any provision of the Clean Air Act (or any regulation thereunder)."

The Clean Air Act Amendments require states to incorporate in their implementation plans permit fees that are to be paid by the owners or operators of major stationary sources. Section 110 states that the fee must be sufficient to cover:

> "(i) the reasonable costs of reviewing and acting upon any application for such a permit, and
>
> (ii) ...the reasonable costs (incurred after such date of enactment) of implementing and enforcing the terms and conditions of any such permit..." (whether the permit was received by the owner before or after the date of enactment),

This provision will no doubt lead to the further application of permit fees based on emission levels to support the costs of administration. The Clean Air Amendments also include cost-based penalties for non-compliance. Non-compliance penalties and permit fees based on emissions could be construed to be emission charges. Non-compliance penalties are designed to punish companies for failure to meet plant-by-plant regulations, while constructed emission charges are incentives to meet ambient air quality standards. Non-compliance penalties and permit fees do not meet our definition of emission charges which encourage cost-efficient abatement.

Non-attainment provisions of the Clean Air Amendments require states to initiate plans to insure attainment in all areas no later than December 31, 1987. This poses the very difficult task of specifying cleanup measures and appropriate technology to be applied to existing as well as new sources. There is a likelihood of misallocation of resources under this provision.

G. Revised Proposal

A joint Ad Hoc Task Force consisting of representatives from both the Pricing Task Force and the Air Quality Task Force was set up to recommend a more focused application of emission charges along with more specific recommendations on which part of the regulations would be replaced. Their report which follows, is entitled "Recommendation of the Emission Charge Ad Hoc Task Force--An Alternative to Individual Source Emission Limitations in State Implementation Plans."

II. Recommendation of the Emission Charge Task Force

A. Background

Regulations have done much in helping to solve our air pollution problems. They have encouraged a substantial abatement of pollutant emissions and an improvement in air quality. Indeed, there are those who argue that no alternative approach can possibly serve us better in the future.

However, despite the progress being made using current methods, questions remain as to the applicability of the present regulatory system for solving all of our air pollution problems. The weighing of costs and benefits is often not done in a manner to facilitate economies in implementation. The judgment of regulators is not infallible. On occasion, resources are misapplied in one problem area while another is not adequately confronted. And, perhaps more importantly, the present system does a poor job of harnessing the innovative forces of our free enterprise system.

If there are shrotcomings to the present regulatory process, what alternative approaches might be acceptable? What problems do we now have that might not yield to the current procedures? These and similar questions led to an examination of a number of current situations to determine whether anything other than traditional regulatory controls might be useful. Particularly, an examination of the many proposals for utilizing pollution abatement charges was made to see if one proposal or a variation might be a useful optional substitute for current regulatory approaches.

Initially, it was necessary to deal with features of pollution abatement charges that have generated the most adverse comment. It is generally agreed upon that no one should be able to buy a "license to pollute." The Task Force could not support any proposal that would have such an effect. Pollution abatement programs should not be designed as vehicles for the purpose of raising revenue for either a local or the federal government. No system of pollution abatement charges should be permitted to become simply an "add on," but rather should be carefully designed to substitute for certain elements of the present system.

The problem of cleaning the air in non-attainment areas is a particularly difficult one for both those who regulate and those who are regulated. Proposals advanced thus far include "trade-offs" which are cumbersome or are of limited utility for small plants and business groups new to an area. The ability of such an approach to preserve competitive forces is in question. The method of enforcing a clean-up or even the shutting down of a plant to effect compliance with a particular State Implementation Plan presents severe difficulties as well.

The proposal which follows is an effort to deal with the problems and constraints associated with regulation as outlined above. It is directed to the solution of problems peculiar to the utilization of coal, and more particularly to the following pollutants: sulfur dioxide, particulates, and oxides of nitrogen generated from the burning of coal.

An effort has been made to present to the Plenary an idea of sufficient detail to permit discussion. Flexibility is built in to reflect regional differences. Final details of the alternative plan will be resolved only by actual trial on an experimental basis. No effort has been made to explore the applicability of pollution abatement charges to other pollutants and regions. If the idea presented by the Task Force survives critical review, the Plenary might wish to consider its use in other situations. For example, the alternative plan may be applicable in other areas where available PSD increments reach zero due to new source additions. The plan developed by the Ad Hoc Task Force is called the "alternative plan" to distinguish it from the State Implementation Plan required by law.

B. Elements of the Proposed Alternative Plan

1. The alternative plan would be available only in non-attainment regions and would apply only to the particular pollutants for which the region is not in compliance. A region could be defined as a multi-state area, a state, or less than a state.

2. No money collected would inure to the benefit of any governmental agency.

3. The alternative plan would be used only if elected by a state as a substitute for those State Implementation Plan (SIP) source emission limitations placed on the industries and pollutants covered by the alternative plan. If a state so elected, after EPA review and approval of the terms and conditions of the plan, it would become part of the SIP, replacing the previous individual source emission limitations for the defined industries, regions, and pollutants.

4. The EPA could not approve an alternative plan that was not designed to bring about compliance with ambient standards in the region for the particular pollutant. The alternative plan would be required to specify an overall emission reduction schedule acceptable to both the state and EPA for the selected pollutant in the region.

5. The alternative plan would apply to a given industry, a group of industries or all industries responsible for a particular pollutant depending on the characteristics of the region. Groupings would be made to ensure competition in pollution abatement. Smaller regions with few companies responsible for a particular pollutant would have different industry groupings than larger regions with more companies. Those industries not included in the alternative plan would be covered by the conventional SIP. Although the alternative plan is aimed primarily at pollutants resulting from coal combustion, other sources of these same pollutants (both large and small) could also be included.

6. The alternative plan would specify a year-to-year emission reduction schedule for each of the pollutants and all of the industries covered by the plan. (Note that the emission reduction schedules would apply to the industries as a whole, not to individual companies or plants.)

7. The alternative plan would specify an initial charge or fee to be paid per unit emissions of pollutant. The initial fee would be set at a level shown through analysis to be adequate in attaining the scheduled emission reduction as emitters acted to minimize their total costs (including capital, operating, and emission charges). Those regions with severe short-term problems due to meteorological conditions could specify time differentiated charges. Surcharges based on three-hour and twenty-four hour averages could be applied during air pollution alerts.

8. All monies collected from each industry group would be distributed annually to the members of that group in proportion to each member's production (of kilowatt hours, tons of steel, etc.). Non-homogeneous industry groupings would require a common denominator for refunds such as total Btu output of heat sources or value added.

9. The difference between actual and scheduled emissions would be reviewed annually. The scheduled reduction in emissions will probably not be a linear function. Rather, there will be stepped reductions as equipment with considerable lead time is designed, built, and put into operation. The charge would be adjusted periodically if actual reductions in emissions differ substantially from planned reductions. If in any year the actual emissions exceeded scheduled emissions, the fee for the next year would automatically increase by a specified percentage (for example, five percent for each one per - cent difference in emissions). The percentage increase would be selected on the basis of its being great enough to result in a rapid reduction of the excess of actual over

scheduled emissions while recognizing the time needed to implement large investment decisions and the desire to provide relatively stable incentives for investment. In regions where application of the system results in pollutant concentrations well below the ambient secondary standards, and where other effects of emissions (such as formation of sulfates, acid precipitation and visibility impairment) are not of concern, the state and EPA may revise the charge downward.

10. The annual review would include an assessment of the question of whether the specified emission reduction schedule is sufficient to meet ambient standards. If it appears to be insufficient, then a more stringent emission reduction schedule would be substituted.

Bibliography: Cogeneration

1. Anderson, Frederick A. et al., Protecting the Environment through Economic Incentives: The Case for Charges, (Washington, D.C.: Johns Hopkins University Press for the Environmental Law Institute and Resources for the Future, forthcoming, 1977).

2. Asmus, Edward Barry, Measuring the Economic Damages from a Proposed Coal-Fired Generating Plant: A Case Study, presentation at the Western Economic Association Meetings held in Anaheim, CA in 1977 (Department of Economics, Boise State University, 1977).

3. Carson, William D., Jr., PhD., On Applying a Sulfur Emissions Tax in Selected Urban Regions, paper prepared for presentation at the 1976 annual meeting of the Western Economics Association,(California Air Resources Board, 1976).

4. Council on Law-Related Studies, Effluent Charges on Air and Water Pollution: A Conference Report, Reporter: Edward I. Selig, Report of Conference held on October 15-16, 1971, (Washington, D.C.: Environmental Law Institute, 1973).

5. Irwin, William A. and Richard A. Liroff, Economic Disincentives for Pollution Control: Legal, Political and Administrative Dimensions, (Washington, D.C.: U.S. Environmental Protection Agency, 1974). EPA contract No. 68-01-2203

6. Johnson, Ralph W. and Gardner M. Brown, Jr., Cleaning Up Europe's Waters: Economics, Management, and Politics, (New York: Praeger, 1976).

7. Kneese, Allen V. and Charles L. Schultze, Pollution, Prices, and Public Policy (Washington,D.C.: The Brookings Institute, 1975).

8. Public Interest Economics, Pollution Charges and Public Policy in California: A Seminar Summary, summary of seminar held on June 24, 1977 in San Francisco, CA, Public Interest Economics (San Francisco, CA, 1977).

9. Smith, Fred Lee, Jr., Pollution Charges - The Applied Literature (Washington, D.C.: EPA, March 1977).

10. Tether, Ivan J., Economic Disincentives for Pollution Control: An Introduction (Environmental Law Institute, 1977).

11. Train, Russell E., Emissions Charges and Other Economic Approaches to Environmental Programs, Tax Notes, Vol. 5, (Feb. 28, 1977).

12. U.S. Congress, Congressional Research Service, Environment and Natural Resources Policy Division, Pollution Taxes, Effluent Charges, and Other Alternatives for Pollution Control, prepared for the Committee on Environment and Public Works of the U.S. Senate at the request of Sen. Muskie (Wash., D.C.: GPO, 1977).

13. Watson, William D., Jr., Control Costs and Damages of Alternative National Pollution Control Policies, Discussion paper D-10, Institutions and Public Decisions Division, Resources for the Future, (Washington, D. C.: May 1977).

14. Yohe, Gary W., Substitution and the Control of Pollution--A Comparison of Effluent Charges and Quantity Standards Under Uncertainty, Albany Discussion Paper Series, (Department of Economics, State University of New York at Albany, January 1976).

SECTION 6
COGENERATION: A BIBLIOGRAPHY

Jerome Kohl, Extension Specialist
Department of Nuclear Engineering
North Carolina State University, Raleigh

and

Judith Dolce, Engineering Reference Librarian
North Carolina State University, Raleigh

Proposed Literature Search
On-Site Power Generation
Significant Words & Phrases

(1) Co-generation

(2) Total energy

(3) Process heat

(4) Process steam

(5) Integrated energy system

(6) Waste heat

(7) Heat recovery

(8) On-site power

(9) On-site generation

(10) Dispersed energy

(11) Dispersed generation

(12) Integrated utility system & modulated integrated utility system (MIUS)

(13) In-plant electric power generation

(14) Combined production of power and heat

(15) Decentralized generation

(16) Advanced technology mix energy systems (ATMES)

(17) By-product electricity

"Can the Fuel Cell Bridge the Gap," Power, Vol. 13, No. 7, (July 1969).

Describes how when a commercial fuel cell becomes available, the all-fuel energy system stands a better chance of competing with the all-electric system.

"Coal Generation Electricity at On-Site Location Being Developed," Aware, (August 1972).

Evaluating Integrated Utility Systems, NTIS Report PB 238765, (September 1974).

The development of utility systems designed to serve the needs of residential complexes ranging from 300-1000 dwelling units is reviewed. Although commercial hardware is available to accomplish separate objectives of an integrated utility system, no system has yet been designed that effectively integrates these functions. Well-planned demonstrations could advance development of integrated utility systems.

"Load Shedding -- Tight Control Spearhead Energy Program," Power, Vol. 20, No. 11, (November 1976), p 53.

The energy management system at FMC's Green River, Wyoming, Soda Ash plant provides integrated computer-assisted control of key functions: load shedding, automatic generation control, real and reactive demand monitoring, voltage regulation, remote control of breakers and top changers, sequence of events recording, boiler monitoring and alarming, automatic synchronizing of generators and ties, and data acquisition and logging. Service interruptions from the local utility justified the sophisticated system. Features of FMC's new system are described. (5 diagrams, 10 photos)

Procedures for Feasibility Analysis and Preliminary Design of Total Energy Systems at Military Facilities, Final Report, (November 1976), 140 pp.,GRAI7706, (Contract: N62399-73-C-0029; Project: DA-4-A-762720-A-896; Task: 4-A-762720-A-89602; Monitor: CERL-TR-E-96).

This manual provides procedural guidance and technical information for performing feasibility analyses and preliminary design of heat recovery installations, especially total energy and selective energy utility plants. This document is intended for use by facility and district engineers and for guidance of headquarters personnel. The feasibility analysis procedure is divided into five major steps: 9L0 identification of gross energy requirements; (2) assessment of the viability of heat recovery within the project environment; (3) assessment of user system constraints and energy requirements -- site selection and analysis, load analysis; (4) def'n. of candidate system configurations; (5) technical and economic eval'n. of candidate systems--energy efficiency analysis and economic analysis. AD-AO33756/8ST

"Starrett City Readies for Startup," Power, Vol. 119, No. 1, (January 1975), p 30.

A large development in Brooklyn, New York depends on its own energy plant for uninterrupted air-conditioning, heating, and electrical service. The physical plant of the total-energy system consists basically of four 875-psig, 110,000-lb/hr steam generators, gas- or oil-fired, that supply steam to two 7.5-Mva turbine-generator sets. Three dual-fuel engines drive generators to supplement the steam-driven sets. Initial electric load is expected to run around 12 mw. Air-conditioning equipment includes 7500 tons of steam-turbine-driven centrifugal water chillers, with extraction steam taken from the turbine-generators. An 8-in. pipleine brings natural gas to the plant and fuel oil is trucked in. Ten days' oil supply can be stored. Present plans call for additional tanks for engine fuel. At some future time, a city sewage-treatment plant across the street may supply methane gas. The plant has been divided down the middle in an effort to prevent complete blackouts.

"Total Energy for Lower Costs," Engineering, Vol. 207, No. 5374 (April 25, 1969).

An analysis of total energy systems using a computer program developed to simulate such systems has shown that they can offer big economies in power production, especially if design is computer-optimized.

"'Total Energy' -- System Plan for the Cascades," Electrical Consultant, Vol. 91, No.3, (March 1975), pp 16-19.

A total energy system feasibility study for the Cascades shopping center in Ocala, Florida contemplated the development of preliminary estimates of cost differences that would exist between a total energy system and a system that would involve the purchase of electricity by each individual tenant with air-conditioning being provided by the owner through a central, electric motor-driven air-conditioning system. With the total energy system as envisioned, the only utility bill would be for natural gas purchased by the developer. The total annual cost of $185,380 for owning and operating the total energy system compares directly to the figure of $344,054 for the purchased power system. The difference, $158,666, represents the estimated savings that would accrue to the total energy system with no difference in cost of the environmental services provided.

"Total Energy Systems and the Gas Turbine Combined Cycle," Electrical Consultant, Vol. 91, No. 3, (March 1975),p.24.

One result of soaring fuel prices and the demand for conservation is renewed interest in the gas turbine combined cycle, which joins a gas turbine and a steam turbine, harnessing the power from each. Combined cycles have proven to be practical in the large horsepower (40,000-horsepower range) gas turbines; however, a study is being made of a combined cycle in which a smaller horsepower gas turbine would be coupled with a boiler, condenser, feedwater pump, valving, and a steam turbine for a power plant suitable for driving equipment including gas compressors, electrical generators and pump.

Achenbach, Paul R. and J. Bert Coble, Application of Total Energy Systems to Housing Developments, No. 361, Vol. 1, Nat'l Bureau of Standards Special Publication (1972), pp 419-430.

A pilot investigation of the performance of a total energy system was undertaken by the National Bureau of Standards to evaluate the potential for decreasing the amount of fuel required to provide the utility services to an apartment complex of 500 units and for better control of noise and air pollution. This installation is being made at a site in Jersey City as a part of the Breakthrough program of the Department of Housing and Urban Development. The pilot total energy plant is being extensively instrumented to determine its thermal efficiency, the daily and seasonal load patterns, the reliability of the utility services, the level of noise and pollution control, the maintenance and repair requirements, the owning and operating costs, and the occupant response to his environment. Five references.

Achenbach, Paul R. and John B. Coble, Site Analysis for the Application of Total Energy Systems to Housing Developments, Intersociety Energy Conversion Engineering Conference,7th, Sept. 25-29, 1972, San Diego, CA, pp 1-31, National Bureau of Standards, Washington, D.C., September, 1972.

In early 1970 a program was approved to design and construct about 2800 dwelling units on eleven sites in the United States with the objective of encouraging the industrialization of the home-building process. As part of this program, operation Breakthrough, a full-scale field study, would determine whether or not total energy systems could provide economical and reliable energy services to apartment complexes and maintain a high level of environmental quality. Characteristics of the eleven sites that were important in determining their suitability for a total energy system were studied as a basis for selection. Fourteen parameters were investigated in the feasibility study. The study resulted in the choice of a city in New Jersey as the preferred location for the pilot installation. COM-74-50619/7

Adam, A.W., and J. Monahan, 100 KW Rankine Cycle Total Energy System, Intersociety Energy Conversion Engineering Conf., 8th, August 13-16, 1973, Philadelphia, PA. Part I, pp. 126-310. Published by Amer. Inst. Aeronautics and Astronautics, NY.

The organic rankine cycle total energy system concept for on-site commercial power up to several hundred kilowatts has been under development since 1968. This paper introduces power selection considerations, discusses working fluid selection, describes the 100 kw breadboard/prototype system and components, and presents performance characteristics. Photographs of hardware and the 100 kw unit on test, plus program status and market potential are included. Two references.

Adam, A. Warren, 100 KW Organic Rankine Cycle Total Energy System Dev't. Status, Conference on Natural Gas Resources and Technology, June 1972, Session II, Part 4, 15 pp. Published by Institute of Gas Technology, Chicago.

The organic rankine cycle total energy system is a new concept in on-site power systems for commercial power requirements of up to several hundred kilowatts. This paper presents the power system considerations, discusses working fluid selection and describes the 100kw breadboard/prototype system and components performance characteristics. Program status and market potentials are also presented.

Agarwal, J.C. and J.R. Sinek, The Energystics of Metal Production, Energy Use and Conserv. in the Metals Ind. Symp. Proc. AIME Annual Mtg.,104th, New York, (Feb. 16-20, 1975) pp. 121-129, Published by Metal Soc. of AIME, NY, 1975.

The primary metals industry has traditionally shown its ability to cope with uncertainties: declining grade of ore, uncertainties inherent in exploration for new deposits, vagaries of a world market characterized by instability of political regimes, and environmental problems. Now a new uncertainty has been added relating to energy supply for industry. The authors have established energy profiles for several base metals. A discussion on the interchangeability of energy forms indicated that it is not always possible. Where possible, cost considerations of the total energy system investment are advised as aids to energy planning in lieu of extrapolation of past fuel costs. Once an energy form has been decided, an adequate trade-off between capital and energy cost must be pinpointed.

Ahuja, Ashok, "Repowering Pays Off for Utility and Industry Plants," Power Engineering, Vol. 80, No. 7, (July 1976), pp 50-54.

Examples of three types of repowering are cited which show that the addition of carefully chosen components to an existing plant to form a combined cycle can increase efficiency as well as capacity while keeping required investment down. These examples concern steam turbine repowering in which gas turbines and new heat recovery boilers are added to an existing steam electric generating plant; boiler repowering in which gas turbines are added to the existing steam generating facilities for power generation, requiring the conversion of existing conventional boilers to heat recovery type boilers; and gas turbine repowering in which a steam generating plant is added to an existing gas turbine plant.

Alich, John A. et al., "Suitability of Low-BTU Gas-Combined-Cycle Electric Power Generation for Intermediate Load Service," Combustion, Vol. 46, No. 10, (April 1975), p 8.

Because system load varies with time, significant portions of system capacity are unused much of the time. Numerous models have been developed to aid in power generating system selection. Model variables associated with the existing system and its demand and alternative power generating systems under consideration are discussed. The generalized concept of coal and oil gasification for electric power production is examined and the economics of low-BTU gas combined-cycle power generation for intermediate load service are compared with alternatives. While economics dictate integration of gasification and power generation for base load service, nonintegrated facilities may be more suitable for intermediate load service from both economic and technical perspectives. Ten diagrams, 2 graphs, 17 references, 2 tables.

American Gas Association: Research and Development, *American Gas Association Report* (1974).

The current American Gas Association Research and Development Program centers on two objectives: new methods for increasing gas supplies and new measures to improve efficiency in gas energy utilization. Association funds for support of the utility research program amounted to $6.6 million in 1974: 51 percent for gas supply, storage, and distribution research; 16 percent for climate control and appliances; 16 percent for basic research and investigation studies; 10 percent for commercial, industrial, and on-site power uses; and 7 percent for environmental protection and enhancement. An additional 2.2 million was supplied by cooperating manufacturers, governmental agencies, and other organizations. One diagram, numerous photos, one table.

Axford, D.G., "Factors in Total-Energy Studies," *Hospital Engineering* (GB), Vol. 27, (February 1973), pp 6-9.

Over recent years, there have been many treatises on total energy, with numerous reports extolling the virtues of an engineering philosophy which can be applied to produce highly attractive returns on capital. Undoubtedly there are countless installations throughout the world where this is true, but the adoption of total-energy principles cannot of itself be relied upon to guarantee economic success. Nevertheless, total energy, properly considered and carefully and appropriately applied, can indeed show handsome returns. One reference.

Barber, R.E., *Potential of Rankine Engines to Produce Power From Waste Heat Streams*, 9th Intersociety Energy Conversion Engineering Conference Proceedings, 1974. (New York: American Society of Mechanical Engineers, August 1974).

This paper describes heat recovery rankine cycle systems and how to estimate the potential power output of a heat source. An economic study shows that the rankine cycle system can produce power at a real cost below that of a coal-fired steam unit and pay off the capital costs in less than three years based on the current minimum value of electricity (2.2 cents/kw hr). This payoff period will be reduced as the value of electricity increases in the future. Seventeen references.

Barber, R.E., Small Rankine Cycle Total Energy System for Recreation Vehicles: A Comparison of Three Possible Approaches, 8th Intersociety Energy Conversion Engineering Conference Proceedings, 1973. (New York: American Institute of Aeronautics and Astronautics, August 1973).

A unitized system which provides one ton (1000 kg) of air-conditioning,20,000 BTU/hr (5000 kg Cal/hr) of heating, and 500 watts of 100 v 60 hz power is estimated to weigh approximately 170 pounds (80 kg) and can be packaged in 10-second, 48-second, 60-second (25-120-150 cm) shells. When providing full air-conditioning and 500 watts of power, the fuel consumption is calculated to be approximately 3 lb/hr (1.4 kg/hr). Typical conventional reaction vehicles presently use a 2.5 kw motor-generator set, a one-ton electrical air conditioner, and a propane heater to provide total energy system functions. In the heating mode, the rankine cycle system utilizes rejected heat from the rankine cycle powered electrical system. This feature can result in significant fuel savings over the conventional system. Three rankine cycle systems were evaluated: (1) two fluid high frequency electrical; (2) low speed conventional electric air-conditioning; and (3) low speed shaft driven air-conditioning. The study shows that system (3) is optimum.

Beckham, W.S., Jr., and F.A. Kuene, MIST (MIUS Integration and Subsystems Test) Laboratory: A Testbed for the MIUS (Modular Integrated Utility System) Program, (American Society of Mechanical Engineers), Paper No. 74-ENAs-44,(August 1974).

The NASA Urban Systems Project Office at JSC awarded a contract to the Hamilton Standard Division of United Aircraft Corporation in mid-1973 to design and construct a testbed integrated utility system called the MIST laboratory. The facility was completed early in 1974 and has begun a long-term program to develop data for MIUS. A description of the

MIST is presented and the critical design issues, which can only be verified by test, are discussed in detail.

Berg, Charles A., *Energy Conservation Through Effective Utilization*, (Washington, D.C.: National Bureau of Standards, Institute for Applied Technology), Report No. NBSIR-73-102, (February 1973), 54 pp. GRAI7315, (Project: NBS-4000910, Monitor: 18).

In two major sectors of the economy, building services and industrial processes, accounting for approximately 75 percent of total national energy consumption, energy utilization is found to be inefficient. It is estimated that in these two sectors, as much as 25 percent of the energy consumed annually by the nation as a whole may be lost through ineffective practices. Possible reasons for existence of ineffective utilization are discussed. Three possible levels of effort to promote effective utilization of energy are identified -- one promotes effective use of present fuels in present processes, the second promotes utilization of presently unused energy sources, the third promotes more effective investment of energy in durable and maintainable products.

COM-73-10856

Bjerklie, J.W., "Small Gas Turbines for Total Energy Systems," *Actual Specifying Engineer*, Vol. 26, No. 2, (August 1971), pp G38-41.

Ways are proposed for using Brayton cycle engines in small total energy systems to conserve fuel and energy. The concepts described do two valuable things for small total energy systems -- they enable the total shaft load to be leveled out for better overall operation and they enable waste heat not demanded by heating or air-conditioning to be put to practical use generating fuel that can be put back into the system or used elsewhere.

Boegly, W.J., *et al.*, *MIUS Technology Evaluation: Collection, Treatment, and Disposal of Liquid Wastes*, NTIS Report ORNL-HUD-MIUS-16, (December 1974).

The modular integrated utility system program conducted by HUD is described. The program is devoted to development and demonstration of the technical, economic, and institutional advantages of integrating the systems for providing all or several of the utility services for a community. The utility services being considered include electric power, heating and cooling, potable water, liquid waste treatment, and solid waste management. An evaluation intended to describe the methods that can be used in a MIUS for the collection and disposal of liquid wastes is presented. The HUD program's goal is to effect, through development and demonstration, early implementation of the integrated utility system concept by the organization, private or public, selected by a given community to provide its utilities.

Bohn, T., Remote Nuclear Heating and Remote Nuclear Energy. NTIS Report AEC-TR-7576, (January 1975). Translation of Report JUL-1077.

The feasibility of using nuclear energy for supplying heat to homes, small users, and industry is discussed. Nuclear energy offers several advantages compared with current, conventional methods in economy, dependability of supplies, and conservation of raw materials. By utilizing waste heat from currant generating processes, remote nuclear heating will prove environmentally preferable to current heating methods.

Brandli, Allan E. and Charles F. Donham, Application of a Numerical Differencing Analyzer Computer Program to a Modular Integrated Utility System, (New York: American Society of Mechanical Engineers), Paper No. 740059, (1974), pp. 547-772.

This paper describes the application of a numerical differencing analyzer computer program to the thermal analyzation of a MIUS model. The MIUS model which was evaluated is one which would be required to support a 648-unit Garden Apartment Complex. This computer program was capable of predicting the thermal performance of this MIUS from the impressed electrical, heating, and cooling loads.

Bresler, Sidney A., Economics of Seawater Desalting in Combination with Ammonia and Power Production, NTIS Research and Development Progress Report PB 203-269, (August 1969), 191 pp, GRAI7122, (Contract: DI-14-01-0001-1485, Monitor: OSW-PR-458).

The feasibility of lowering the operating cost of producing desalted water by combining its production with ammonia and electricity was examined. The cost of producing desalted water from sea water in a single-purpose multistage flash distillation (MSF) plant can be reduced almost 30 percent by combining its production with that of ammonia, to about 27-35 cents per thousand gallons in plants of capacities ranging from 12.5-180 MGD. The use of a common energy center permits the design of plants having a wide range of relative capacities of ammonia, desalted water and/or electric power, resulting in further possible reductions of unit water costs. The use of nuclear energy starts to become economically attractive when the cost of natural gas or other hydrocarbon fuel exceeds 25 cents per million Btu.

Brown, Seymour W., "Selective Energy -- A System That Splits Load and Saves Money," Actual Specifying Engineer. Vol. 32, No. 2, (August 1974), pp 84-86.

While studying the feasibility of using a total energy system for a college campus, the engineers found that a selective energy system incorporates the best features of total energy at a lower cost.

Buderath, J., "Part Played by RWE (Rheinish-Westfalischen Elektrizitatswerks Ag) in the Nordrhein-Westfalen Energy Centre," Energiewirtsch Tagesfragen, Vol. 26, No. 1-2, 5-12, (January - February 1976).

Discusses the importance in the Nordrhein-Westfalen region of the Essen/Gelsenkirchen/Muhleim area with its high bounded by Cologne, Aachen, and Dusseldorf with its soft coal deposits and the large RWE power stations. Reference is made to the historical development of RWE and the important part played by coal in electric power generation. Involvement in nuclear power generation and work carried out by the company in the fields of heat recovery, heat exchangers, heat pumps, solar energy and dual electrical/thermal heating schemes. Four references.

Burchnall, A., "Private Generation and Combined Heat/Power Schemes," Electrical Review, Vol. 199, No. 4, (July 23, 1976), p. 12.

The British Government's policy on nationalized industries is to treat them as commercial bodies with the understanding that investment options should be placed where the rate of return to the national economy is greatest, but difficulties have developed in coordinating heat and power schemes between private and public industries. Statistics are provided for rate differentials for both the private and the public sectors. Two photos, one table.

Buskirk, B.A. and A.R. Spiewak, Lists of Environmental Statements Projects, June 1975, 44pp. (Contract: W-7405-eng-26, Monitor: 18).

The document is a collection of five lists of the Environmental Statements Projects related to nuclear power reactors, consumer products in which by-product radionuclides may be used, ERDA facilities, and, for the Department of Housing and Urban

Development, Modular Integrated Utilities Systems. The lists are arranged by (1) project name, (2) project docket number, (3) NRC Environmental Project Manager, (4) ORNL Task Group Leader, and (5) ORNL editor. In each list, the first issue date listed for each project is the issue date of its Draft Environmental Statement (DES) and the second issue date is that of the Final Environmental Statement (FES). The lists will be updated as the Environmental Statements Project continues. NSA3210; NTIS Report # is ORNL-TM-4985

Carter, K.L., "CO Waste Heat Recovery Can Pay for Itself," Hydrocarbon Processing - Petroleum Refiner, Vol. 42, No. 9, (Sept. 1963).

Castenschiold, R., "Multiple Utility Services and On-Site Generation," Electrical Construction and Maintenance, Vol. 75, No. 1, (January 1976), pp 57-59

Here are the most important points to consider in providing dual energy sources, including the vital function of switching from one to the other.

Christensen, Laurits R. and William H. Greene, "Economies of Scale in U.S. Electric Power Generation," Journal of Political Economy, Vol. 84, No. 4, (August 1976), p 655.

Economies of scale for U.S. firms producing electric power are estimated. Cross-section data for 1955 and 1970 are analyzed using the translog cost function. Significant scale economies were available to nearly all firms in 1955. But by 1970 the bulk of U.S. electricity generation was by firms operating in the essentially flat area of the average cost curve. A small number of extremely large firms is not required for efficient production, and policies designed to promote competition in electricity generation cannot be faulted in terms of sacrificing economies of scale. Three graphs, 11 references, 10 tables.

Christiansen, Alden G., Waste Heat Utilization/Reduction, EPA National Conference on Health, Environmental Effects, and Control Technology of Energy Use, Feb. 1976. Published by EPA Industrial Environmental Research Lab in Cincinnati.

Waste heat rejection from U.S. central power stations by the year 2000 will nearly equal the total energy consumed in 1970. Use or reduction of waste heat offers numerous environmental and energy benefits. EPA activities in waste heat utilization and reduction applications are described. Research and development efforts focus on three options: by-product waste heat utilization, by-product electrical generation by industry, and integrated energy production/use complexes. By-product waste heat utilization has the greatest potential for near-term energy savings. Six references, one table.

Claffey, C.J. and C.L. Segaser, MIUS Technology Evaluation: Compression Refrigeration Systems for Air Conditioning, Energy Research Abstracts 0106, (February 1976), 46 pp, (Contract: W-7405-eng-26; Monitor: 18).

The technology of compression refrigeration for air-conditioning systems is well established. Its widespread application in warmer climates throughout the world attests to its general acceptance as a reliable means of providing comfort air-conditioning. Currently, a diverse selection of air-conditioning equipment is available to suit any air-conditioning application. With the exception of some of the room air-conditioners, compression machines produce the desired cooling effect in an efficient manner. Some of the small capacity machines, however, are notably efficient, thus proving that a relatively high efficiency is attainable even in the smaller machines. This evaluation describes the current status (as of May 1973), performance, and costs of the types of compression refrigeration equipment considered suitable for use in MIUS to provide residential space cooling. The following topics are discussed in this report: (1) the fundamental refrigeration cycle; (2) a variety of types of mechanical refrigeration compressors; (3) selection of refrigerant fluid; (4) efficiencies of different systems; (5) performance values applicable to different systems; (6) energy flow; and (7) capital, operation, and maintenance costs. In addition, a listing of manufacturers of compression-type air-conditioning equipment and their products is provided. Final selection of component size and type depends on the design of the MIUS installation in which the refrigeration system would be included and on the cooling requirements of the community to be served. Life-cycle cost analyses, which should be a major factor in the selection of any of the MIUS subsystems, require systems analyses based on the particular MIUS design, the community to be served, and other site-specific characteristics. Such systems and life-cycle cost analyses are outside the scope of this report. NTIS Report ORNL-HUD-MIUS-19

Coble, John B. and Paul R. Achenbach, "Description of Equipment and Instrumentation for a Field Study of a Total Energy System in an Apartment Development," _Proceedings of the Seventh Intersociety Energy Conversion Engineering Conference_, (Washington, D.C.: National Bureau of Standards, Sept. 1972), 27 pp., GRAI7418, (Project: NBS-4628422; Monitor: 18).

The Breakthrough site at Jersey City, NJ is the location for an installation and field study of a total energy system. This development covers six acres and is comprised of four buildings containing 488 dwelling units, a small building for commercial use, two small schools, a swimming pool, and the total energy plant. The field study is aimed to produce much-needed authoritative information on engineering performance, maintenance requirements, and load-and-cost data for total energy systems. The central plant and the individual buildings are being extensively intrumented to provide digital data on fuel utilization, the generation of electrical and thermal energy, the excess heat rejected, and the utilization of electrical and thermal energy by all major segments of the load. COM-74-50614/8

Coble, John B., _et al_., _Performance of the Engine-Generators Used in the Jersey City Total Energy Plant_, Vol. 11, HUD Utilities Demonstration Series, (Wash., D.C.: Nat'l Bureau of Standards, Final Report No. NBSIR-77-1207, (Oct. 1976) 22 pp., GRAI7711, (Mon.:18).

Each of five 600-kilowatt (kw) diesel engine-generators which were to be installed in a total energy plant was performance tested under NBS direction at the engine-generator vendor's plant. These tests provided a basis for acceptance of the engine-generators and for comparison with installed performance. This testing was performed as a part of a comprehensive study to assess engineering, economic, and environmental aspects of a total energy plant which supplied all electrical power, hot water, and chilled water to an apartment complex in Jersey City, NJ. Under sponsorship of the Dept. of Housing and Urban Development (HUD), the Nat'l. Bureau of Standards (NBS) has instrumented the total energy site for engineering data and is collecting economic and environmental data. The engines were tested at seven power levels ranging from 0 to 110 percent rated electrical load. In the tests, fuel consumption, electrical output, and jacket-water heat recovery were measured as well as many other parameters. Results are reported for fuel consumption, electrical efficiency, and electrical-plus-thermal efficiency and comparisons are made with measured data from the total energy plant PB-264 427/6ST

Coble, John B. and Paul R. Achenbach, Site Analysis and Field Instrumentation for an Apartment Application of a Total Energy Plant, (Washington, D.C.: National Bureau of Standards, Department of Housing and Urban Development), Final Report No. NBSIR-75-711, (May 19, 1975), 65 pp, GRAI7517 (Project No.: NBS-4628422, Monitor: 18).

Under sponsorship of the Department of Housing and Urban Development, the National Bureau of Standards developed criteria in a feasibility study to select a site for, and evaluate the requirements of, a total energy system on one or more Operation Breakthrough housing sites. The total energy system produces its own electrical, heating, and cooling energy services independent of the local utility system. A Jersey City site was chosen as the location for the installation, evaluation, and field study of the total energy system. COM-75-10689/8ST

Compere, A.L., et al., MIUS Technology Evaluation: Water Supply and Treatment, (Energy Research and Development Administration, April 1976), 148 pp, GRAI7625, (Contract: W-7405-eng-26; Monitor: 18).

Water treatment is one of the services which may be performed within a Modular Integrated Utility System (MIUS). A MIUS can serve from 100 to 3000 dwelling units and their associated facilities. The small size and self-contained nature of a MIUS create special situations with respect to the application of innovative water treatment and distribution technologies. This report evaluates current and future technology on the basis of its suitability for application as part of a MIUS development. A MIUS may include water treatment for domestic use or fire protection. The conventional technology evaluated in this report is, for the most part, appropriate to the treatment of surface water supplies and the small pre-engineered plants described are designed primarily for surface water treatment. In some situations, separate components, rather than a complete pre-engineered plant may be most appropriate when designing a MIUS supply. A MIUS must provide adequate disinfection as a part of the treatment process, and in some cases a desalting process is necessary. The cost of water treatment will vary as a function of water source quality and use. In order to insure adequate plant operations, personnel must be provided in accordance with appropriate regulatory agency standards. Some automation may be possible, but it is doubtful that all manpower can be eliminated. Improper operation or maintenance can result in water unfit for human consumption or other MIUS uses. (ERA Citation 01:016295) ORNL/HUD/MIUS-21

Collet, P.J., "Energy Saving by a Combination of Total Energy Systems and Heat Pumps," Ingenieur (Netherlands), Vol. 86, No. 19, (May 9, 1974), pp 358-362.

A block diagram is given showing the use of a heat pump for more effective space heating than is achieved with the present method. Forecasts are made of the sources of the energy to be used in the Netherlands in 1985: 5 percent nuclear, 3.3 percent coal, and the remainder natural gas and oil. By use of the heat pump principle a saving of 16.8 percent is estimated. It is proposed to make all this saving on oil. Nine references.

Crawford, J.J., "Total Energy -- A Realistic Answer to Fuel Conservation," Electron Power, Vol. 19, No. 10, (May 31, 1973), pp 210-212.

Potentially useful heat is wasted on an enormous scale by, for example, using cooling towers at power stations to condense steam. The article argues that "total energy" systems can offer economic advantages as well as conserving natural resources.

Crawley, E.F. and J. Jewell, Application of the Medium Speed Diesel Engine for Generation of Electrical Power, (Diesel Engineers & Users Assn cation), Report No. 373, (May 1976), 19 pp.

The paper presents a broad survey of the advantages, scope, economy, and versatility of diesel power for permanent or temporary electrical generating stations. Essential features for automatic starting and operations are described. Loading and environmental factors have a large bearing upon the size and number of generating sets installed, the type of installation, and the ancillary equipment necessary. These are discussed in detail with the requirements of cooling systems and heat recovery arrangement. The essentials of maintenance and length of operational periods before overhauls are described and tabulated. Three references.

Daglish, A.G. and J. Rich, "Total Energy for a Modern Hospital Complex (Total Energy Fuer Ein Modernes Krankenhaus)," Gas Waerme International, Vol. 19, No. 9, (Sept. 1970), pp 350-355.

The paper outlines the concept of on-site generation of electricity from natural gas with waste-heat recovery in hospitals. A total energy plant is described for a modern hospital in which the energy requirements for the complex are supplied by gas. Recovered heat from the prime movers is used, together with supplementary heat when required, for hospital heating and cooling services and for distribution to outlying buildings. The cost of such a scheme is compared with that for conventional purchased power and direct-fired boilers.

Dain, J.J. and D. Whitlock, "Optimization of Total Energy Systems," British Chemical Engineering, Vol. 14, No. 9, (Sept. 1969), pp 1209-1212.

The ideal total energy system is one in which the maximum energy is extracted from the fuel by means of a careful balance and sizing of the system components with due consideration for capital cost. Power Gas Corporation Ltd. has developed a computer program which simulates a total energy system and gives the ideal component sizes for optimum running or capital costs. Its use enables an instant assessment of the effect on cost of design modifications or variations in the energy requirements. The program results include highly detailed technical and financial analyses of the chosen system and, thus, also allows a clearer perspective of the essential design requirements.

Decker, Gerald L., Power Plant Design Considerations for Energy Conservation, (Chicago: Illinois Institute of Technology), American Power Conference, (April 1976).

Demetroulis, Nicholas M., Technical Review and Analysis of the Total Utility Demonstration Plant Design and Operational Concept, Final Report, (September 1976), 55 pp, (Contract: DAAK02-75-0096; Monitor: USAFESA-RT-2031).

The total utilities concept involves the furnishing of all utilities to a facility from a single plant as a means of more efficiently using available energy from our fuel sources. These utilities would consist of all or part of the following: electricity, heating and air-conditioning, water supply, sewage treatment, domestic hot water, and

trash disposal. The current practice is to obtain most of the above services from separate, non-affiliated corporate entities whose plants are widely separated and who have little incentive for cooperative action in waste energy utilization. Most electric generating plants, for example, are of large capacity and are remotely located so that even though vast waste energy quantities are available, the cost for collection, storage, and distribution of such energy to consumers is prohibitive. Great amounts of energy are dissipated through cooling water, stack gases, and power transmission which result in the loss of up to 60 percent of the energy initially available in the fuel. Another obvious resource waste is the practice of trash disposal through open burning, land fill or incineration without extraction and use of the considerable heat energy available.
AD-A037 016/3ST

Denton, Jesse C., et al., Energy Conservation through Effective Energy Utilization, (Washington, D.C.: National Bureau of Standards, National Science Foundation; New York: Engineering Foundation), Report No. NBS-SP-403, (June 1976), GRAI7621, (Monitor: 18), 254 pp.

Effective utilization of energy and other natural resources; national benefits of energy conserfation; options for energy conservation; measures of thermal energy utilization; industrial energy analysis and forecasting; management's role in industrial thermal energy utilization; a projection of energy demand by the iron and steel industry; potential energy savings in the forming of ventilating and air-conditioning equipment; factors controlling the manufacture and marketing of energy conserving products; financial feasibility of energy conservation. PB-255-878/1ST

Doherty, Martin, C., "Gas Turbine's Role in LNG Liquefaction and Vaporization," Pipeline Industry, Vol. 42, No. 3, (March 1975), p 57.

Heavy duty two-shaft gas turbines will be a good fit in many future LNG liquefaction plants. Heat sources and safety and efficiency in LNG vaporization terminal plants are described. The use of gas turine in-plant generation with heat recovery for LNG vaporization is discussed. One diagram.

*<u>Energy Industrial Center Study</u>, (Washington, D.C.: National Science Foundation; Midland, MI: Dow Chemical Company; Ann Arbor, MI: Environmental Research Institute of Michigan; New York: Townsend-Greenspan and Company, Inc., Cravath, Swain and Moore), Final Report, (June 1975), 472 pp , GRAI7522, (Grant: NSF-DEP74-20242; Monitor: 18).

This report summarizes the findings of a study of the technical, environmental, economic, legal, and regulatory aspects of two well-known approaches to conserving fuel burned to produce steam: (1) the generation of by-product electric power from the steam generated by industry for use in manufacturing, and (2) the supplying of steam to industry from utility central power stations. Estimates are developed of the nationwide extent to which each approach is economically viable and of the corresponding reductions in capital investment and fuel consumption. An assessment is made of the potential impacts on the electric utility industry and on electric rates over the next decade. Environmental issues to be faced in specific instances of implementation are identified for the pulp and paper, petroleum refining, and chemical industries. Attention is given to possible interactions with the Federal Power Act, the Public Utility Holding Company Act, state regulatory patterns, antitrust laws, environmental protection considerations, fuel allocation priorities, private law considerations, and federal income tax consequences. PB-243 823/2ST

Dubin, Fred S., <u>et al</u>., "Guidelines for Saving Energy in Existing Buildings," <u>Engineers, Architects, and Operators Manual, ECM 2</u>, (Washington, D.C.: Federal Energy Administration, Office of Energy Conservation and Environment), (June 16, 1975), 434 pp, GRAI7610, (Contract: DI-14-01-0001-1844; Monitor: FEA/D-CP-21).

This report is intended for engineers, architects, and skilled building operators who are responsible for analyzing, devising, and implementing comprehensive energy conservation programs. It includes energy conservation measures which can result in further energy savings of 15 to 20 percent with an investment cost that can be recovered within 10 years through lower operating expenses. PB-249 929/1ST

*Important recent paper.

Dubin, F.S., "Total Energy Systems for Mass Housing," (New York: American Institute of Aeronautics and Astronautics), *Second Urban Technology Conference and Display*, (July 1972), San Francisco, CA.

The author describes a program to develop a screening manual which readily indicates the relative economic feasibility of total energy systems for mass housing projects. This manual would also include graphs for determining the relative impact of alternate systems on the environment.

Dubin, Fred S., "Total Energy for Mass Housing -- Why It Makes Sense," *Actual Specifying Engineer*, Vol. 29, No. 2, (February 1973), pp 69-82.

The author explains when it is in the public interest to use decentralized power plants to supply the base electrical load and, through waste heat recovery, most of the energy for heating and cooling mass housing projects. 18 references.

Dubin, Fred S., "Total Energy Systems and the Environment," *Actual Specifying Engineer*, Vol. 28, No. 4, (October 1972), pp 57-64.

The author explains why it is becoming increasingly important to seek new sources of energy and build decentralized total energy systems. He discusses thermal and system efficiencies, total energy and central station fuel requirements, and emission factors in boilers and prime movers. 41 references.

Eller, Willard M., "Power Generation in Cane Sugar Factories," *Combustion*, Vol. 46, No. 3, (September 1974), pp 17-26.

The article illustrates the means by which power production may be increased, such as, by the use of higher boiler operating pressures and temperatures, high efficiency boilers, more efficient steam cycles, heat recovery methods, and the use of modern instrumentation to maintain stability and ease of operation. Several examples show how the efficiency may be increased.

Elliott, Thomas C., "Second Generation SO_2 Scrubbing System Leads the Way for Industrial Power Plants," *Power*, Vol. 118, No. 11, (November 1974), p 26.

Pioneering efforts by General Motors at the Chevrolet-Parma plant near Cleveland, Ohio, resulted in the development and implementation of an advanced system of sulfur dioxide removal from industrial power plant stacks. The advanced double alkali system now nearing completion is described. One diagram, two photos, three tables.

Ferden, S.L., et al., Energy Systems Optimization Computer Program (ESOP) Developed for Modular Integrated Utility Systems (MIUS) Analysis, Intersociety Energy Conversion Engineers Conference, Paper No. 749058, (New York: American Society of Mechanical Engineers), August 1974, pp 541-546.

This paper presents a description of a computer program, the Energy Systems Optimization Program (ESOP) developed jointly by Lockheed Electronics Company (LEC) and the Urban Systems Project Office of the National Aeronautics and Space Administration/Johnson Space Center. This program predicts the loads, energy requirements, equipment sizes, and life cycle costs of alternative methods of meeting these utility requirements. The program has been used extensively for performing energy analyses of Modular Integrated Utility Systems (MIUS), a concept which is being sponsored by the Department of Housing and Urban Development. 21 ref.

Fraas, A.P., et al., Design Study for a Coal-Fueled Closed Cycle Gas Turbine System for MIUS Applications, Intersociety of Energy Conversion Engineers Conference, 10th, (New York: Institute of Electrical and Electronics Engineers, Catalog No. 75CHO 983-7 TAB), Paper No. 759042, (August 1975), pp 260-268.

A firm design for a pilot plant has been evolved from a feasibility study of the use of a fluidized bed coal combustion system coupled to a closed cycle gas turbine for modular-integrated utility system applications after considering a wide variety of possibilities and their relative merits in meeting the basic requirements. Ten references.

Fraas, A.P., Small Coal Burning Gas Turbine for Modular Integrated Utility Systems, Intersociety Energy Conversion Engineering Conference, Paper No. 749017 (New York: American Society of Mechanical Engineers),(August 1974),pp 499-507.

The increasingly difficult supply situation for gaseous and liquid fuels and for electricity introduces a new element of risk into construction and operation of large new residential complexes. One logical approach for both the owners of the complex and the nation as a whole is to develop and employ a total energy system that will operate on coal. Such a system would supply electricity, hot water, building heat, and air-conditioning using 4 to 8 units of 350 to 500 kw(e) to provide excess capacity and redundancy for high reliability. To meet environmental standards, this system must include provisions for keeping sulfur emissions to acceptable levels, and this implies the use of a fluidized bed coal combustion chamber which makes it possible to remove about 90 percent of the sulfur in the coal by combining it with crushed limestone in the bed to give calcium sulfate. The paper summarizes a series of design studies, outlines the principle possibilities, and indicates the effects of major parameters on both the capital cost and the thermal efficiency. 13 refs.

Fraas, A.P. et al., Use of Coal and Coal-Derived Fuels in Total Energy Systems for MIUS Applications, Vol.1, Summary Report, (April 1976), 48pp, GRAI7626, (Contract: W-7405-eng-26; Monitor: 18).

Various ways in which coal and coal-derived fuels might be employed in MIUS systems have been investigated. The fuels considered include high and low sulfur bituminous coals, lignite, anthracite, the products of various coal gasification, liquefaction, and solvent refining processes, coal loaded with sewage sludge after being used as a filtering medium, and coal mixed with residential and industrial solid wastes. The types of power conversion systems considered for use with these fuels include gas engine, diesel engine, conventional gas turbine, open and closed cycle gas turbine with fluidized bed furnace, steam turbine and engines with conventional furnaces, and steam turbine and engines with fluidized bed furances. The principal conclusions are: (1) MIUS systems based on small, on-site coal gasification or liquefaction plants would have much higher capital and operating costs than systems in which coal is burned directly; (2) a fluidized bed combustion system in which coal is burned in a bed of limestone appears to be the most attractive system for direct utilization of coal -- sulfur emissions in the stack gas are reduced by a factor of about ten, and any type of coal or lignite or solid waste can be burned; (3) a closed cycle gas turbine coupled to a fluidized bed coal combustion system appears to be the most promising system for using coal for MIUS applications. This system will convert about 30 percent of the energy in the fuel to electricity and about 50 percent into heat that can be used for domestic hot water and building heating and air-conditioning. (ERA citation 01:018582). ORNL/HUD/MIUS-27.

Fraas, A.P., et al., Use of Coal and Coal-Derived Fuels in Total Energy Systems for MIUS Applications, Vol. 2, (April 1976), 352 pp, GRAI7626, (Contract: W-7405-eng-26; Monitor: 18).

Detailed engineering and feasibility studies on the MIUS system are reported. This system consists of fluidized-bed combustion of coal and/or various other fuels or wastes with limestone particles in the bed to absorb the sulfur with fuel. The application was limited to power for 500 to 1000 residential units so gas turbines appeared to be more suitable in such small sizes than steam turbines and, after considerable study, a closed cycle system was chosen. The studies leading to this overall system and its optimization and control systems are given. (ERA citation 01:018583) ORNL/HUD/MIUS-28

Frei, D.H., "Gas Turbine -- As a Total Energy System for Paper Mills," Pulp and Paper International, Vol. 16, No. 10, (Sept. 1974), pp 48-54.

Efficient heat and energy generation and utilization have become all the more critical at today's increased fuel prices. The gas turbine is presented here as a potential prime mover for total paper mill energy supply systems -- producing electricity plus steam and direct hot exhaust gas for drying. The gas turbine system appears particularly well-suited for tissue mills. Overall operating cost savings and compact installation are two advantages. Technical and operational details of a paper and pulp mill-based gas turbine power station are included.

Frei, D.H. and H. Holik, "Total Energy Supply Systems for Paper Mills," Combustion, Vol. 45, No. 10, (April 1974), pp 39-43.

The production of paper, cardboard, and millboard requires energy in three forms as a rule: electricity, steam, and hot gas for drying. The present article describes a total energy system capable of generating these three energy components simultaneously in a compact unit. The economics are shown taking the consumption figures for a tissue machine as an example. In addition, the economical application area is delimited as a function of the ratio between power and heat consumption.

Frei, D.H. and H. Holik, "Total Energy Supply Systems for Paper Mills," Sulzer Technical Review, Vol. 55, No. 3, (1973), pp 189-194.

The present article describes a total energy system capable of generating three energy components simultaneously in a compact unit. The economics are shown taking the consumption figures for a tissue machine as an example. In addition, the economical application area is delimited as a function of the ratio between power and heat consumption.

Fulbright, B.E., MIUS Community Conceptual Design Study, (Houston: National Aeronautics and Space Adminstration, Lyndon B. Johnson Space Center), Report No.: NASA-TM-X-58174, (June 1976), 565 pp, (Monitor: 18). STAR1423

The feasibility, practicality, and applicability of the modular integrated utility systems (MIUS) concept to a satellite new-community development with a population of approximately 100,000 were analyzed. Two MIUS design options, the 29-MIUS-unit and the 8-MIUS-unit, facilities were considered. Each resulted in considerable resource savings when compared to a conventional utility system. Economic analyses indicated that the total cash outlay and operations and maintenance costs for these two options were considerably less than for a conventional system. Computer analyses performed in support of this study provided corroborative data for the study group. An environmental impact assessment was performed to determine whether the MIUS meets or will meet necessary environmental standards. The MIUS can provide improved efficiency in the conservation of natural resources while not adversely affecting the physical environment. N76-33085/1ST

Gamze, M.G., A Critical Look at Total Energy Systems and Equipment, (Washington, D.C.: American Chemical Society), Proceedings of the Seventh Intersociety Energy Conversion Engineering Conference, (1972)

The author defines the total energy concept, reviews the history of the concept, and describes the problems involved in designing and installing the systems. The basic problem is that such systems have to be custom designed around components that are not always totally suited for the application. This necessitates extensive engineering to design a dependable installation and rigorous maintenance procedures to insure system reliability after the job has been placed in operation. In addition, the author presents a detailed listing of the problems of the concept. This is followed by specific recommendations to improve upon existing hardware and to develop new hardware in order to make the total energy concept a viable one which will permit its potential for conserving fossil fuels and reducing air, water, and thermal pollution to be realized.

Gamze, M.G., "Total Energy System Design Primer, I" Intersociety Energy Conversion Engineering Conference, 10th, Aug 18-22, 1975, Newark, Delaware. pp 456-60.
Based on the discussion of the impact of the user systems, i.e., heating, cooling and electrical requirements, This section deals with the electrical load profiles as encountered in real buildings.

Gamze, Maurice G.,"Total Energy System Design Primer, Continuation," Intersociety Energy Conversion Engineering Conference, 10th, Aug 18-22, 1975, Newark, Delaware. (New York: Institute of Electrical and Electronic Engineers, Catalog No. 75CHO 983-7 TAB), Paper No. 759072, (August 1975) pp 456-460.
Same as preceeding reference.

Gamze, Maurice G.,"Total Energy Systems Operation," Intersociety Energy Conversion Engineers Conference, Paper No. 769290, 1976, pp 1683-1687. (New York: American Institute of Chemical Engineers)
Part of the operation of a total energy plant is described with the purpose of indicating a maintenance philosophy, historical data, and provide a guide for future maintenance at both the specific plant and projected plants.

Garay, P.N., "Symbiosis to Reduce Energy Crunch," Chemical Engineering, Vol. 81, No. 20, (Sept. 30, 1974), pp 126-127.

Gerstin, R.A., Technical and Economic Overview of the Benefits of Repowering,(American Society of Mechanical Engineers), Paper No. 75-GT-16, (March 1975), 8 pp.
General consideration for matching combustion and steam turbines are discussed as a background for the typical repowering study. Different arrangements of the combustion turbines and heat recovery boilers are presented, discussed, evaluated, and a comparison of the systems are made. To complete the investigation, a series of economic factors are considered for comparison between repowering and an alternate form of generation. In conclusion, by incorporation of combustion turbines in repowering of an old steam plant, the extended use of combustion turbines can be justified.

Gerstin, R.A., "Technical and Economic Overview of the Benefits of Repowering," Combustion, Vol. 46, No. 10, (April 1975), pp 17-22.

General considerations for matching combustion and steam turbines are discussed as a background for the typical repowering study. Different arrangements of the combustion turbines and heat recovery boilers are presented, discussed, evaluated, and a comparison of the systems are made. To complete the investigation, a series of economic factors is considered for comparison between repowering and an alternate form of generation. In conclusion, by incorporation of combustion turbines in repowering of an old steam plant, the extended use of combustion turbines can be justified.

Gervais, Robert L. and Piet B. Bos, "Solar Thermal Electric Power," Astronaut Aeronaut, Vol. 13, No. 11, (Nov. 1975), pp 38-45.

Solar thermal conversion (STC) systems collect solar radiation and convert it to heat in a working fluid for use either in a thermal-electric conversion system or in a total energy system which delivers both electrical and thermal energy. Two generic types of STC power-plant concepts exist, namely, central-receiver systems and distributed systems, the distinction being that in a central-receiver system solar energy is transmitted optically from collectors mounted on two-directional tracking systems to a tower-top receiver (or boiler) near the turbogenerator, while in a distributed system, numerous solar energy collectors are connected by long piping runs to transport heated fluid to a turbogenerator. Alternate types of solar energy collectors include flat-plate collectors, which can use total insolation (direct plus diffuse) and focusing types such as parabolic-trough or paraboloidal-dish collectors which can only utilize direct insolation. Thus, flat-plate collectors, or augmented flat-plate collectors including side plates, work reasonably well on hazy or cloudy days. However, they are relatively inefficient for power production, although it is conceivable that future mass production cost savings or low-cost system innovations could offset their inefficiency. Tentative projections for solar thermal electric conversion and solar total energy systems are 40,000 mw/e and 8,000 mw/t, respectively, by the year 2000. Ultimately, solar thermal conversion systems could be expected to supply 20 to 30 percent of the nation's electrical energy and 10 to 20 percent of the nation's thermal energy for residential, commercial, and industrial needs. Earliest commercial application of solar thermal conversion systems, consistent with this program, is expected to be in the 1985-1990 time period.

Gill, W.L., et al., Technology Evaluation of Heating, Ventilation, and Air Conditioning for MIUS Application, (Houston: National Aeronautics and Space Administration, Lyndon B. Johnson Space Center), Report No. NASA-TM-X-58128, JSC-08571, (Nov., 1974), 63 pp, (Monitor: 18). STAR 1301

Potential ways of providing heating, ventilation, and air conditioning for a building complex serviced by a modular integrated utility system (MIUS) are examined. Literature surveys were conducted to investigate both conventional and unusual systems to serve this purpose. The advantages and disadvantages of the systems most compatible with MIUS are discussed. N75-10135/2ST

Glaser, C.W. and J.D. Morgan, Building Heating-Cooling-Ventilating Energy Analysis Computer Program, IEEE Conference Record of 1971 Industrial and Commercial Power Systems and Electric Space Heating and Air Conditioning Joint Technical Conference, 1971, pp 164-168. (New York: Institute of Electrical and Electronic Engineers),(May 1971).

The purpose of the program is to provide a reasonably quick and accurate method for evaluating heating-cooling ventilating systems and their energy input requirements. This paper outlines the mathematical formulation and methodology used in writing this program. The program is written in Fortran IV and originally was loaded for use on an IBM 1130/8K computer, but has since been converted for use with the IBM 360/50 computer. The 360 version requires about 75 K of core and 300 K of direct access working files.

Glenn, D., et al., Economic Evaluation of Total Energy, Technical Report No. HUD-DSC-2, (July 1973), 490 pp, GRAI7410, (Contract: HUD-H-1776; Monitor: 18).

Total energy systems are on-site power generation systems with heat recovery. By generating electricity on the premises, it is possible to recover normally wasted heat and use it to provide space heating, cooling, and domestic water heating. The objective of the study was to develop economic evaluation quidelines which could be used by the private and public sectors as a screening tool to determine the applicability of total energy systems for residential developments. PB-228 893/4

Glenn, D., et al., Economic Evaluation of Total Energy: Guidelines, Report No. HUD-DSC-1, (July 1973) 59 pp, GRAI7410, (Contract: HUD-H-1776; Monitor: 18).

These guidelines provide a screening tool which can be used by the private and public sectors to determine the economic applicability of total energy systems for residential developments. The guidelines are designed to allow a simple, straightforward, low-cost determination of whether or not a complete engineering feasibility study is justified. They do not eliminate the need for a complete feasibility study requiring the services of engineering consultants experienced with total energy systems. PB-228 683/9

Glenn, Donald R., Housing Data for Modular Integrated Utility System Economic Analysis, Vol. II, (Washington, D.C.: Department of Housing and Urban Development, Policy Development and Research), Report No. HUD-DSC-3-Vol-2, (May 1974), 236 pp, GRAI7517, (Contract: HUD-H-1776; Monitor: 18).

The study examines the multifamily housing construction industry. Emphasis is placed on the present market and the growth patterns displayed during the last five years when this industry experienced its greatest advance. During this period, the multifamily construction market share increased from 40 percent to almost 54 percent of total new housing. The final report is presented in three volumes. This volume presents the multifamily project characteristics, including average number of buildings per project, number of living units and stories per building, cost per square foot as a function of number of stories per building, and other financial relationships relative to land use as well as average living unit size per project. These averages are compiled by census region and by climatic regions defined by cooling degree day isopleths starting at 500 and increasing in increments of 500 to 4500, resulting in 10 zones. The 46 SMSA's leading the nation in multifamily housing construction are overlaid onto the climatic zones. PB-242 393/7ST

Goen, Richard L., Assessment of Total Energy Systems for the Department of Defense, Vol. I, (Stanford Research Institute), Report No. SRI-EGU-2513-Vol-1, (November 1973), 122 pp, GRAI7418, (Contract: DACA23-73-C-0014, ARPA Order-2408; Monitor: 18).

The purpose of the study is to assess the potential applicability of various types of total energy systems to military installations. The types of energy systems considered include diesel, gas turbine, steam turbine, geothermal, solar, nuclear, and solid wastes. Fuel savings are given for each type of system and their costs are compared with the costs of conventional systems. The two most promising systems are solar energy applied to heating and cooling and nuclear power. AD-781 816/4

Goen, Richard L., et al., Assessment of Total Energy Systems for the Department of Defense, Vol. II, (Stanford Research Institute), Final Report No. SRI-EGU-2513-Vol-2, (Nov. 1973), 160 pp, GRAI7418, (Contract: DACA23-73-C-0014, ARPA Order-2408; Monitor: 18).

The purpose of the study is to assess the potential applicability of various types of total energy systems to military installations. The appendix volume of the final report contains: (1) engineering performance characteristics and costs of fossil fuel system elements; (2) energy consumption data for military bases and derivation of the energy load profiles used in the study; (3) description of the fuel consumption model and summaries of the fuel consumption and total system costs for the various cases; (4) characteristics and costs of the geothermal systems; and (5) description of solar energy systems. AD-781 817/2

Granitzer, H., "Total Energy Power Stations -- A From of Energy Supply Favorable to the Environment," Energie (Germany), Vol. 26, No. 12, (Dec 1974), pp 431-435.

Considers the importance of noise generated in power stations by generators, circulating and boiler feed pumps, electric motor drives, safety valves, coal handling plant, and cooling towers. The effect of rating, speed, and design features on the noise of air-conditioning motors is considered and noise reduction methods to meet specific requirements are discussed. Reference is made to the Klagenfurt total energy power station. The advantages of total energy schemes with waste heat recovery and the economics of heat energy transport costs are briefly considered.

Guiliani, P., "Combined Production of Heat, Cold, and Electric Current at the Centre International de Paris," Revue Generale de Thermique (France), Vol. 12, No. 134, (February 1973), pp 141-153.

The energy requirements of the Paris International Center amount to 28,000,000 KCAL/hr, 12,000,-000 frigories per hour, 20,000 KVA, of which 5000 KVA are being produced independently. The author explains the different solutions considered for this power production and then describes the solution retained, i.e., low pressure, hot water boiler-house, production of frigories by electric centrifugal machines, and continuous production of part of the electric power requirements by generators driven by thermal motors of the dual-fuel type with heat recovery in the form of hot water. He notes the main characteristics of this combined production. This achievement uses different power supply sources: electric current, natural gas for feeding the dual fuel motors with pilot injection of fuel, and natural gas and fuel for feeding the compound burners of the boilers. The generating sets run continuously and have a capacity of five mw.

Hafele, W. and W. Sassin, Applications of Nuclear Power Other Than for Electricity Generation, Proceedings of the European Nuclear Conference: Nuclear Energy Maturity, (April 1975), pp 107-119. Vol. 2, (New York: Pergamon Press).

Primary energy shares and the end use of energy are considered first since applications of nuclear power other than for electricity generation have to fit into these constraints. The most immediate application is to various industries. A more detailed consideration of the growing importance of secondary energy as a consumer-oriented highly specialized energy form points to gaseous fuels. For expediting the substitution of fossil primary energy for heating purposes, hydrogen produced electrolytically from offpeak electric power of nuclear stations, and its possible market share are considered. The truly large scale H^2 production via thermolysis as a key to a mainly nuclear-based energy scenario leads to the long-term aspects of the other applications of nuclear energy. The deployment of large-scale fuel cycles is considered for an asymptotic scenario where nuclear fuels meet the entire primary energy demand. This enhances the aspect of fuel cycle colocations and their siting. Here artificial islands are of interest. The reasoning culminates in decision trees for the deployment of nuclear power in particular, and advanced energy systems in general. 12 references.

Harmsworth, John Peter,"Basic Economics of Industrial 'Total Energy' Systems," Vol. 6, World Energy Conference, 8th, Paper No. 3, (Bucharest: Romanian National Committee of the World Energy Conference),(July 1971). 17 pp.

The phrases "total energy" and "on-site" power systems are relatively new promotional names for the old concept of generating electricity or shaft horsepower at or near the point of utilization. On-site generation implies that the heat released by the prime mover is not used, whereas total energy systems use some or all of the available heat. In both cases, the main purpose of the system is to produce electrical or shaft power. The primary incentive to use total energy systems lies in the potential savings to the user. There are many possible ways to supply the heat and power needs of a factory and the selection of the optimum method is a complex problem. A method of visualizing the relative importance of some of the major factors in the selection is proposed and an illustrative example is included. Reference is also made to the services provided by third-party companies and the likely effect of total energy systems on energy policies. It is concluded that the current compromise of centrally generated power and locally generated heat can be both wasteful of fuel and capital resources in comparison with a good total energy system.

Harmsworth, John Peter, "An Introduction to Total Energy Systems," Vol. 1, Proceedings of the Conference on Total Energy, (London: Institute of Fuel), (December 1971).

The main considerations affecting total energy systems are briefly covered to act as an introduction to the subject. The interest in total energy systems lies in their potential for reducing costs of power and heat to industrial and commercial users. The three main types of systems, using back-pressure steam turbines, gas turbine, and reciprocating internal combustion engines are briefly described and a simple method of visualizing the source of the potential energy bill savings from each system is indicated. The need to make detailed analyses of the various methods of supplying the heat and power requirements of each plant is emphasized. Comment is also made on the subjects of reliability, availability, type of power required, design and maintenance of total energy systems, their financing, and their contribution to reducing atmospheric pollution and improving the national economy.

Harvey, D.G., *et al.*, *Urban Energy Systems: Session II*, (San Diego: Seventh Intersociety of Energy Conversion Engineers Conference, September 1972), pp 1247-1269.

Papers contained herein include: *Minimization of Residential Energy Consumption, Off-Peak Air-Conditioning Using Thermal Energy Storage, Critical Look at Total Energy Systems and Equipment*, and *Role of the New York City Government in Energy Conservation*.

Hise, E.C., *et al.*, *MIUS System Analysis: Comparison of MIUS and Conventional Utility Systems for an Existing Development*, (Energy Research and Development Administration, June 1976), 178 pp, GRAI7707, (Contract: W-7405-eng-26; Monitor: 18).

This study was directed toward the application of the Modular Integrated Utility System (MIUS) concept to an existing housing-commercial development (Walden) that is located near Chicago, Illinois. The postulated MIUS (the live model) would supply electricity, district heating and cooling, liquid waste treatment, and solid waste disposal to the development. Potable water for human consumption and fire protection were not included in the study. Earlier MIUS modeling efforts were made on models that allowed simpler analysis, but were not representative of actual housing developments. Comparisons of fuel savings, annual costs, and environmental impacts between the MIUS and conventional utility services to the development have been performed. In the live model, even though the source of electricity was small (engine-generators having an operating efficiency averaging about 30 percent), the overall fuel utilization was about 62 percent. The overall efficiency of fuel utilization for the conventional utility was only 51 percent. Economic analyses using present marginal cost for incremental services (as of January 1972) indicate that MIUS costs more than would be required to serve the same development conventionally. Environmental impacts of the live model are an increase in CO, HC, and NO/sub x emissions, but a reduction in SO/sub 2 and particulate emissions. In any case, the live model will not exceed the Federal Air Quality Standards (FAQS) and the local air quality is not markedly changed from that of the conventional utilities. Impact due to on-site liquid waste treatment is mainly caused by releasing effluent at a different point than a large conventional system since the same or higher degree of treatment can be obtained with small plants as can be produced with larger plants. (ERA citation 02:005169) ORNL/HUD/MIUS-20

Hittle, Douglas C., Total Energy and Total Utility Systems for Conservation of Resources, (Champaign, IL: Construction Engineering Research Lab), Interim Report No. CERL-IR-E-61, (March 1976), 33 pp, GRAI-7612, (Project: DA-4-A-76720-A-896; Task: 4-A-76720-A-89602; Monitor: 18).

This report describes the problems encountered in the feasibility assessment and design of total energy/total utility systems, outlines the general framework of a computational model for solving these problems, and describes the data required to validate such a model. AD-A023 244/7ST

Holcomb, R.S., Flow Visualization Studies of a Fluidized Bed Furnace for a Coal-Fired Closed Cycle Gas Turbine, (New York: American Society of Mechanical Engineers), Paper No. 75-WA/Ener-4, (December 1975), 7 pp.

Flow visualization studies were performed on a small-scale plexiglass model of a coal-burning fluidized bed furnace that has been designed as a heat source for a closed cycle gas turbine total energy system for the HUD-MIUS program. The test model was 10 inches square with a tube bundle of 1/4 inch O.D. tubes six rows deep. The bed material was limestone with a mean particle size of 370 mu and a static bed height of one foot. The following results were obtained: a static region of bed material can be used as thermal insulation of the bed plate and around the turbine air manifolds; the minimum velocity for solids mixing at the wall was 1.75 feet per second, allowing a 3:1 turndown of the furnace; the bed pulsation frequency was two to three cycles per second over the operating range of air velocity. Five references.

Hufford, Paul E.,"Applying Current Technology to Help Alleviate Energy Problems," Intersociety Energy Conversion Engineering Conference, (August 1975), p 247.

The effective exploitation of energy sources already available can help to alleviate the U.S. energy crunch. Solid waste energy and the integrated micro utility system, which is the supply of all the energy requirements for a single site from a central on-site source, can help to buy time until other technologies can be developed.

Hundemann, Audrey S., Total Energy Systems for Buildings (Citations From the Engineering Index Data Base), (Springfield, VA: National Technical Information Service, April 1976), 60 pp, GRAI7611, (Monitor: 18).

Technology for total energy systems as applied to residential complexes and industrial plants is discussed. These systems employ a primary source of energy, such as oil, natural gas or solar heat to provide comprehensive energy requirements in the form of light, heating, cooling, air-conditioning drying, process heat, and power for an industrial plant or for a commercial or public building. Included are schemes for electricity generation, waste heat recovery, and direct power drive of mechanical equipment coupled to a prime mover. Modular Integrated Utility Systems (MIUS) are coverd, including studies on gas turbines for use in MIUS systems, and computerized simulation studies. NTIS/PS-76/0275/8ST

Hundemann, Audrey S., Waste Heat Utilization (Citations from the NTIS Data Base),(Springfield, VA: National Technical Information Service, April 1976), 170 pp, GRAI7611, (Monitor: 18).

Techniques of recovering waste heat from power plants, buildings, industrial processes, and waste disposal are covered. Included are studies of waste heat boilers and the use of waste heat for irrigation, sewage treatment, odor control, desalination, heating, and aquaculture.
NTIS/PS-76/0276/6ST

Hundemann, Audrey S., Waste Heat Utilization (Citations from the Engineering Index Data Base), Vol. 1, (Springfield, VA: National Technical Information Service, April 1976), 184 pp, GRAI7611, (Monitor: 18).

Waste heat recovery from industrial processes, electric power plants, buildings, and incineration is covered, with emphasis on energy conservation in industrial plants. Included are studies of waste heat boilers and the use of waste heat for irrigation, sewage treatment, odor control, desalination, and heating. NTIS/PS-76/0277/4ST

Hundemann, Audrey S., Waste Heat Utilization, (Citations from the Engineering Index Data Base Vol. 2, (Springfield, VA: National Technical Information Service, April 1976), 76 pp, GRAI7611, (Monitor: 18).

Recovery of waste heat from industrial process plants, electric power plants, buildings, and incineration of wastes is covered. Emphasis is on energy conservation in industrial plants. Included are studies of waste heat boilers and the use of waste heat for irrigation, sewage treatment, odor control, desalination, and heating. NTIS/PS-76/0278/2ST

Jacobs, Madeleine, "MIUS May Provide Key to Increased Housing Production," Dimensions-NBS, Vol. 60, No.11, (November 1976), p 3.

A Modular Integrated Utility System is one that brings together several subsystems which serve the different utility needs of a community, and that builds them into a master system which performs the combined functions as well as or better than a conventional system. A MIUS can be used for fuel conservation through recovery of energy that normall is wasted when essential utility services are supplied from separate sources. NBS is currently acting as technical program advisor for the first MIUS demonstration project to be funded by HUD. The MIUS can conserve fuel and other natural resources, reduce environmental degradation, reduce total public costs, match the reliability of service of separate utility systems, and permit greater land movement flexibility. The NBS-HUD program is described. One diagram, four photos.

Jaspers, H.A. and F.K. DuPre, Stirling Engine Design Studies of an Underwater Power System and a Total Energy System, (New York: American Institute of Aeronautics and Astronautics, August 1973).

Future scientific and commercial sea floor exploration will produce a growing demand for underwater power systems. Stirling engines are especially well suited for this application because of their ability to run on non-conventional fuels, their low maintenance characteristics, and the fact that they can be designed for high efficiency. The recent advances in the technology of high temperature heat pipes is of great importance in applications of this nature. An example of such a design is presented. Another important application area might be a total energy system for a home, since it would be capable of producing heating, cooling, and electric power with low fuel consumption. An integrated design of a Stirling engine and a Stirling refrigerator, using any kind of conventional fuel is discussed along with its performance characteristics. Seven references.

Jaumotte, A.L., "Application of Small Gas Turbines to the Production of Power," Energie (Belgium), Vol. 197, No. 4, (1971), pp 191-206.

The author reviews applications of small gas turbines (less than 500 kw) in aeronautical and industrial fields. Advantages and disadvantages in comparison with the diesel engine are discussed. The use of small turbines for combined electrical and thermal energy production (total energy system) is illustrated by thermodynamic characteristics and industrial examples. The article ends with a review of the prospects for gas turbines in aerospace research, notably for spacecraft power supply requirements. 24 references.

Jeffs, E., "More Diesel Generation Could Further Fossil Fuel Economy," Energy International, Vol. 13, No. 5, (May 1975), p 26.

With the launch of seahorse medium-speed diesel engine and an associated heat recovery system, Hawthorn Leslie has embarked on a campaign to secure support for diesel power generation in the United Kingdom as a way of exploiting North Sea oil more efficiently.

Jones, Ronald A., Caterpillar on Total Energy, (Diesel Engineer Users Association), Publ. No. 345, (January 1972), pp 3-16.

Recovery of maximum usable fraction of fuel energy from diesel or natural gas prime movers by electric power, mechanical power, and heat or air-conditioning is discussed. Heat recovery systems for packing plants, motels, schools, stores, and others are outlined.

Kaiserset, Karel, "Ausnutzung der Abdampfwaerme Von Dampfkraftprozessen Durch Die Waermepumpe" ("Waste Heat Recovery from Steam Power Processes with the Heat Pump"), Brennst Waerme Kraft (Germany), Vol. 27, No. 5, (May 1975) pp 219-224.

The processes of steam power generation and heat pump are compared with the help of Carnot-cycle, especially by comparing the back-pressure turbine and the combination of condensation turbine and heat pump. Equations and charts are used to show which possibilities of waste heat recovery exist in the course of the year and under which conditions the heat pump can be usefully employed. 7 references.

Kaplan, E. and E. Lodwig, Design and Performance of a Mobile Waste Heat Organic Rankine Cycle System Providing Electric Power and Environmental Control, Paper No. 699056, Intersociety of Energy Conversion Engineers Conference, 4th, (September 1969), pp 484-493.

An organic Rankine cycle power system has been developed which absorbs heat from the exhaust gas of a 30 kw gas turbine generator producing an additional 18.9 kw which serves to support electrical loads greater than the gas turbine rated capacity and to power an air-conditioning system. Heat rejected from the system provides water and space heating and humidification. Descriptive material and prototype performance test data are included.

Kaplan, E. and E. Lodwig, Organic Rankine Cycle Power System for Waste Heat Reclamation in the Tipi Total Environment Facility, Twenty-third Annual Power Sources Conference, (May 1969), pp 159-162.

Single Tactile Image Processing and Interpretation (TIPI) complex may include a group of Total Environmental Facilities (TEF), which are self-sufficient ground and air transportable lightweight shelters of honeycomb sandwich wall construction, capable of satisfying all electrical demands, maintaining stringent environmental conditions internally, and providing auxiliary support services. The major elements of the Total Energy System are a gas turbine generator set, an organic Rankine cycle system deriving heat input from gas turbine exhaust, an air-conditioner, a water heater, and an air heating and humidifaction system.

Kauffman, W.M., "Consider an In-Plant Engine Clinic to Develop Troubleshooting Ability," Power, Vol. 120, No. 9, (September 1976), p 77.

On-site generator systems have increased in number in recent years, going into new commercial centers,shopping malls, remotely controlled standby power, and isolated industrial complexes. The problem is the limited supply of qualified technicians to maintain the units. The article shows how many operators capable in routine duties can benefit from training sessions complete enough to include problem-solving in crash situations. Programs of two-day training sessions are outlined.

Kellock, B.C., "Petbow Total Energy System," Machinery and Production Engineering (Great Britain), Vol. 128, No. 3302, (March 24, 1976), pp 266-270.

The electrical system of a factory which produces electrical machines is described. As builders and suppliers of private generation plant, the firm planned the electrical system for a new factory for the possibility of it becoming self-contained with regard to power requirements.

Kenkare, A.S., "Economics of Total Energy Systems," Plant Engineering (Great Britain), Vol. 19, No. 12, (January 1976), p 20.

Kennedy, D., et al., Alternative Strategies for Optimizing Energy Supply, Distribution, and Consumption Systems on Naval Bases -- Assessment of Total Energy System Applications at Naval Facilities, Vol. 3, (Bethesda, MD: Booz-Allen and Hamilton,Inc.) Final Report, (November 1974), 146 pp, GRAI7506, (Contract: N62399-73-C-0029; Monitor: CEL-CR-75.003).

The key topics investigated and discussed are advanced research in total energy systems, opportunities for heat recovery from prime movers at naval facilities, and feasibility of using inport steaming to provide power for naval shore facilities. AD/A-003 590/7ST

Kirn, H., "Systems for the Rational Use and Recovery of Energy," Elektrizitaeswirtschaft (Germany), Vol. 74, No. 5, (March 1975), pp 126-129.

Approximately 76 percent of the total useful energy generated in the Federal Republic of Germany is thermal,out of which approximately 40 percent is used for space and 36 percent for process heating. Therefore, the rational use of energy is especially to be obtained through methods aiming at the wise use of thermal energy, its recovery, and its generation from sources which have only rarely been called upon. The author explains the processes applied for the recovery of heat and the generation of energy from environmental heat by means of the heat pump. Seven references.

Klass, D.L., Relationship of the Energy Inputs to the Net Energy Production of Fully Integrated Energy-Producing Systems,(New York: American Chemical Society, Division of Fuel Chemistry Preparation, April 1976), pp 1-7.

The net energetics and the energy inputs into integrated, synthetic energy-producing systems are extremely important to the development of new energy supplies. Basically, the ultimate goal is to design and operate environmentally acceptable systems to produce new supplies of salable energy, whether they be low-BTU gas, substitute natural gas (SNG), synthetic crude oil, methanol, ethanol, hydrogen or electric power from primary raw materials such as coal, oil shale, biomass, organic wastes, and isotopes, at lowest possible cost and with the minimum consumption of energy inputs. It is essential to quantify how much energy is expended and how much salable energy is produced in each fully integrated system. In this paper a basic concept is proposed is which is believed to be broadly applicable and useful for the development of new synfuel supplies. The concept also suggests ground rules for the analysis of the net energetics of fully integrated systems. Support for the methodology is expected from its application to real systems. 11 refs.

Klooster, H.J., "Integrated Process Plants Demand Total Energy Concept at Design Stage," Oil and Gas Journal, Vol. 71, No. 26, (June 25, 1973), pp 93-96.

Larinoff, Michael W., Economic Analysis of On-Site Power Installations, Southeastern Electric Exchange Sales Conference, (November 29, 1965).

Kolb, J.O., Effects on Air Quality from MIUS and Conventional Thermal-Electric Systems, Intersociety Energy Conversion Engineers Conference, 10th, Newark, Delaware, Aug 18-22, 1975. Paper No. 759043, 1975, pp 269-277. (New York: Institute of Electrical and Electronic Engineers, Catalog No: 75CHO 983-7 TAB).

The objective of this study is to evaluate the effects on local air quality from combustion emissions of the thermal-electric subsystem of a Modular Integrated Utility System (MIUS) and several conventional utility system models. The consumer model in this study is a garden apartment complex housing 2500 people. 7 references.

Kurtz, N.D., "Keep It Simple -- Total Energy on the Plains," Actual Specifying Engineer, Vol. 25, No.5, (May 1971), pp 88-91.

A Kansas beef processing facility's demands for hot water and power, plus a need for simplicity, made total energy highly feasible.

Lehmann, Edward J., Energy Conservation and Waste Heat Utilization, (Springfield, VA: National Technical Information Service), Report No. NTIS-WIN-74-051, (July 1974), 147 pp, GRAI7416, (Monitor: 18).

The bibliography contains 142 selected abstracts of research reports retrieved using the NTIS on-line search system. The report is divided into two sections dealing with energy conservation. In the first, 50 reports are presented which concern all aspects of energy conservation including topics such as reducing electricity demand, conservation policies, and the reduction of heating and automotive fuel consumption. The second section covers all aspects of recovering waste heat from power plants, buildings, and industrial water. These 90 reports include studies of total energy systems, waste heat boilers, and the use of power plant waste heat for irrigation, heating, sewage treatment, and desalination. COM-74-11138/6

Lehmann, Edward J., Waste Heat Utilization, (Springfield, VA: National Technical Information Service, February 1975), 122 pp, GRAI7506, (Monitor: 18).

The techniques of recovering waste heat from power plants, buildings, industrial water, and waste disposal are covered in this bibliography containing 116 abstracts. Included are studies of total energy systems, MIUS technology, waste heat boilers, and the use of power plant waste heat for irrigation, sewage treatment, desalination, and heating. NTIS/PS-75/215/4ST

Leighton, Gerald S., Urban Energy Utilization and Conservation, (Chicago: Information Service, Institute of Gas Technology), Paper No. 2, Conference on Natural Gas Resources and Technology, 2nd, June 1972, 13 pp.

HUD's responsibility for improving the present and future well being of existing, and yet-to-be-built urban areas requires that we pay strict attention to energy sources and their efficient use. The objective of HUD's research, which presently covers the spectrum from an analyst of individual dwelling unit energy utilization through the modular integrated utility system, is to assure that the quantities and types of energy that are required to satisfy current and future population needs are available. This can be accomplished by the evaluation of existing and developing technologies and by working to affect institutional change. Three references.

Leijendeckers, P.H.H., et al., The First 420-Bed Total Energy Hospital, Vol. 1, Conference on Total Energy, December 1971. (London: Institute of Fuel), pp 263-278.

The new construction of the Sint Anna Ziekenhuis in Geldrop, completed in 1971, has a completely self-contained on-site power generation installation. The building is suitable for 420 beds, complete with operating theaters, x-ray rooms, polyclinic, etc., and is fully air-conditioned. A total energy plant is designed for four ebullient cooled gas engines, each driving a 700 kva brushless generator. Heat recovery from engine jacket water and exhaust supplies steam for the absorption type air-conditioning and winter heating. Additional steam is supplied by four gas-fired boilers. The electric automatic 220/380 V switchgear and control system automatically starts, stops, and synchronizes the gas engine generators in a predetermined sequence, and in response to electric demand load sensing. The main switchboard and control panel is extensively described.

Limaye, Dilip R., et al.,"Energy Policy Evaluation: Modeling and Simulation Approaches," Workshop on Modeling and Simulation for Energy Policy Evaluation, Nov 1974, p 215. (Lexington, MA: Lexington Books).

The book includes 14 papers describing some of the recent applications of modeling and simulation approaches to energy policy evaluation. The papers cover a broad spectrum of topics and address different levels of decision making, ranging from strategic and tactical issues to operational issues.

Littler, J.G.F., Autonomous House Experiment, (New York: Pergamon Press, 1976), pp 609-620.

Project relying on on-site resources of energy supply at an acceptable price is described. A prototype house with own water supply, waste disposal, wind generator, and solar heating system is proposed and calculations are presented. 10 refs.

Maag, W.L. and G. Bollenbacher, Energy and Economic Analysis of Total Energy Systems for Residential and Commercial Buildings, (Cleveland: National Aeronautics and Space Administration), Report No. NASA-TM-X-71617, E-8099, (Oct 1974), 19 pp, (Monitor:18).

Energy and economic analyses were performed for an on-site power plant with waste heat recovery. The results show that for any specific application there is a characteristic power conversion efficiency that minimizes fuel consumption and that efficiencies greater than this do not significantly improve fuel consumption. This type of power plant appears to be a reasonably attractive investment if higher fuel costs continue. N74-34542/2ST

McCracken, Paul W., et al., Energy Industrial Center Study: Executive Committee Summary Report and Policy Proposals, (June 1975), 23 pp, GRAI7522, (Grant No.: NSF-OEP74-20242; Monitor: 18).

This report summarizes the findings of a study of the technical, environmental, economic, legal, and regulatory aspects of two well-known approaches to conserving fuel burned to produce steam: (1) the generation of by-product electric power from the steam generated by industry for use in manufacturing, and (2) the supplying of steam to industry from utility central power stations.
PB-243 824/OST

McCracken, Paul W., et al., Energy Industrial Center Study, NTIS Report No. PB-243 823-4, (June 1975).

Presented is a study of the technical, environmental, economic, legal, and regulatory aspects of two well-known approaches to conserving fuel burned to produce steam: the generation of by-project electric power from the steam generated by industry for use in manufacturing; and the supplying of steam to industry from utility central power stations. The nationwide extent to which each approach is economically viable is estimated and the potential impacts on the electric utility industry and on electricity rates over the next decade are considered. Environmental issues to be faced in specific instances of implementation are identified for the pulp and paper, petroleum refining, and chemical industries.

Marks, R.H., "Comeback for On-Site Power Generation?" AACE Bulletin, Vol. 7, No. 1, (March 1965), pp 18-26.

Martini, W.R., et al., Stirling Cycle Engines, Intersociety Energy Conversion Engineering Conference, (August 1975), p 995.

Various types of Stirling engine cycles and factors in their performance are discussed. Topics include design and analysis of the free-displacer, free-piston Stirling engine, efficient thermomechanical electricity generation from the heat of radioisotopes, critical length ratio in the Stirling engine heat exchanger, use of solar energy and solid fuel heat sources in a Stirling engine, a lithium/sulfurhexafluoride heat source for an underwater Striling engine, and a lithium/sodium/sulfurhexafluoride heat source with Stirling engine as a submersible propulsion system. 30 diagrams, 16 graphs, 7 photos, 48 references, 9 tables.

Mauro, Robert S., "Reliability and Research Boost Small Generating Units," Public Power, Vol.34, No.2, (March-April 1976), p 10.

The newer forms of small-sized generating units, fuel cells, diesel bottoming cycle, Stirling engine, and advanced gas turbines, all have three important advantages over moder large-scale thermal plants -- they can use various fuels, they require little or no colling water, and they have minimal environmental impact compared with larger generating plants. The Stirling engine and improvements in gas turbines are described. Fuel cells and the use of a bottoming cycle to recover waste heat from diesel engines are discussed. Four photos.

Mayo, W.E. and J.C. Purcupile, A Selective Total Energy System for a Residential Complex, Intersociety Energy Conversion Engineering Conference Proceedings, 8th, 1973, pp 389-395. (New York: American Institute of Aeronautics and Astronautics, 1973).

A modular integrated utilities package is being investigated as a possibility for small, out-of-the-way communities. One of the principle components of this package is a selective total energy system which will supply all of the thermal needs of the complex, but only a portion of the electrical needs. The remainder of the electric energy requirements would come from the central utilities group. As long as the balance between electric demand and thermal demand in the complex exists, the selective total energy plant supplies both electricity and heat, but whenever the electric demand exceeds the thermal demand, which is very often the case, the electric grid of the public utility automatically takes over and supplies what is needed. 8 refs.

Mayo, D.S., et al., Total Energy Conference, Vol. 1, Institute of Fuel, Total Energy Conference, (Dec. 1971), p 372.

Meador, J.T., MIUS Technology Evaluation: Comparison of Trenching Costs for Separate and Common Trenching, (February 1976), 71 pp, (Contract: W-7405-en-26; Monitor: 18).

Estimates are made of the combined costs associated with excavating, backfilling, and compacting several classes of soil in trenches required for the installation of utilities. Earthwork costs incude estimates for both earthmoving equipment and operating-labor requirements. Published cost rates are applied to appropriate sections of trenches in order to estimate total earthwork costs. Trench cross sections used comply with Occupational Safety and Health Administration (OSHA) standards. The derived trenching cost data are then used for estimating the possibility of cost reduction through the application of "common trenching" methods to several configurations of one section of mains that are of sufficient size to meet the demands of a large garden apartment complex served by a Modular Integrated Utility System (MIUS). The results show that earthwork costs could be spacing between utilities rather than be providing separate trenches for each type of utility. However, close spacing would lead to limited accessibility for maintenance, and the add'tl. cost resulting from limited accessibility is not estimated in this study. Potential savings were small in installations that have large spacing between utility service mains thus providing ready accessibility for maintenance. ORNL-HUD-MIUS-15

Meador, J.T., MIUS Technology Evaluation: Thermal Energy Conveyance, (May 1976), 95 pp, GRAI7626, (Contract: W-7405-eng-26; Monitor: 18).

Thermal energy produced by a MIUS can be distributed at moderate temperatures, and low-pressure steam or water are most adaptable as energy-transfer media. This report discusses the types, cost and performance of serveral types of conduits for thermal energy conveyance. Conduits applicable to water conveyance of thermal energy produced in a MIUS are evaluated from data on characteristics and economic factors related to district heating and cooling systems for housing developments. Materials of construction are considered according to the demands of a conduit: (1) the pipe must meet the requirements for conveyance of the heat-transfer media; (2) the insulation must limit thermal losses; and (3) some form of encasement must protect the pipe and insulation from both external loads and the underground environment. Factors such as heat-transfer characteristics, thermal expansion, and creep strength at operating temperature must be considered in the selection of conduit materials. Commercially available prefabricated conduits with or without insulation are considered with respect to their ability to meet energy conveyance requirements at installed costs that are based on present-day economics. Illustrative analyses that include most of the design parameters involved in thermal conveyance are presented to put into perspective the most important features of such systems. Another factor that may influence the selection is the installation labor costs. Prefabricated conduits made of mild steel with corrosion-resistant coatings or of fiberglass-reinforced epoxy seem most likely to meet the requirements of a MIUS for low-temperature hot water (LTHW) service. Conduits made of asbestos-cement or a thermoplastic such as polyvinyl chloride are considered less expensive and adequate for chilled water service. (ERA citation 01:019449)

ORNL/HUD/MIUS-22

Meckler, Milton, "On-Site Energy Systems Stage a Comeback," Electrical Consultant, Vol. 91, No. 3, (March 1975), pp 20-23.

This country is experiencing a significant resurgence of interest in on-site systems because such systems reuse "waste" heat. They about double the efficiency of energy recovery from the fuel source (from typically 35 percent at a conventional electric utility, to about 70 percent for an on-site system). Also, in spite of the high initial cost, a properly designed, installed, and operated system can be paid off quite quickly. Air pollution is reduced, as is water pollution, and the danger of brown-outs and black-outs is eliminated.

Meckler, Milton, "Options for On-Site Power," Power, Vol. 120, No. 3, (March 1976), p. 33.

An approach to choosing the optimum combination of on-site electric power generation and purchased power for a plant is presented. Seven alternative systems and their efficiencies for different load conditions are described. Past failures in system design resulted from lack of expertise and problems with subsequent expansion. Two diagrams, two tables.

Meckler, Milton, "Preparing for On-Site Power <u>Power</u> Vol. 119, No. 4, (April 1975), pp 82-83.

Under the right conditions, on-site power generation can save both money and increasingly scarce fuel. These savings accrue because of inherent restrictions on the conversion of fuel into power for conventional end uses. Under present technology, less than 40 percent of the energy in fuel (and often much less) can be converted into power, either to generate electricity or to drive mechanical equipment. The remaining energy, in the form of waste heat, is ordinarily thrown away. By using the waste heat in processes other than power generation, however, the amount of fuel energy put to useful purposes can be increased substantially. It is also possible to install energy-consuming equipment that relies on heat instead of power -- for example, absorption cooling instead of mechanical cooling. Such an arrangement can have a double advantage: simultaneously reducing overall power requirements and increasing the application of waste heat. A further advantage of on-site generation can be improvement in power quality -- fewer outages and/or more precise control of electrical characteristics, which is often needed for electronic data processing. While on-site generation with heat recovering can lead to substantial cost and energy savings, these savings can be realized only under the proper combination of conditions. A thorough analysis is therefore essential to determine whether or not such a combination exists for a specific facility.

Medici, M., <u>et al.</u>, Development of Heat Recovery Plants in Italy in the Five-Year Period 1962-1967, Vol. 19, Paper No. 37, <u>Seventh World Energy Conference</u>, (August 1968), pp 131-149.

The review includes steam plants with partial and total heat recovery (of regulated offtakes of steam and backpressure), and those of specific, advantageous applications in plants with gas turbine engines, in which the production and distribution of heat is combined with the generation of electrical energy. The paper deals especially with developments in the food, paper, and textile industries.

Meyer, Charles F. and David K. Todd, "Heat-Storage Wells," <u>Water Well Journal</u>, Vol. 27, No. 10, (October 1973), pp 35-41.

The analysis described indicates that large quantities of useful heat may be stored underground in special water wells, with more than three-fourths of the heat being recoverable after 90 days. The economics appear favorable. The hydrological and heat-transfer considerations of heat-storage wells, the markets for heat, and possible total energy system configurations are discussed together with problems of energy conservation and thermal pollution.

Meyer, Charles F. and Wlater Hausz, "New Concept in Electric Generation and Energy Storage," *Annual Power Technology Conference*, 8th, (October 1975), 17 pp. (Stillwater: Oklahoma State University) Paper No.17.

The problem of storage of large amounts of heat for periods of several months, at low cost and low loss, in multi-gigawatt total energy systems -- integrated energy utilities -- is considered. A solution is proposed by advancing the concept of the heat storage well. It allows storage of large amounts of useful heat for 90 days or more, at low cost and with low loss, in an environmentally acceptable way. To be useful for space heating, absorption-cycle air-conditioning, water heating, and process heat, temperatures of 280 to 350 degrees farenheit are suitable. Saturation pressure at 350 degrees farenheit is 135 psia. An ordinary water well is drilled into a confined aquifer -- a porous subsurface layer of sand and gravel overlain by a relatively impervious material such as clay. The combination of lower density and viscosity causes the hot water to override the colder water. After 90 days of injecting one million gallons of water per day into an aquifer formation 100 feet thick, the heat is concentrated in a volume about 500 feet in diameter. For a saturation pressure of 135 psi, a heat of 312 feet is required and the depth to the confining layer should be 400 feet or more. After five or more cycles, the energy loss during the injection-withdrawal cycle is less than 20 percent. The cost of 90-day storage would add only 6 to 10 cents per million BTU to the cost of delivered heat. 24 references.

Mitchell, Edward J. and Peter R. Chaffetz, *Toward Economy in Electric Power*, (American Enterprise Institute for Public Policy Research, National Energy Study), (December 1975) 38 pp.

The conflict between lower prices and shortages and between higher prices and adequate supplies of electric power can be resolved by allowing private industry to utilize excess steam and other energy from its plants to produce electricity at substantial fuel and capital savings. Technical, industrial, economic, financial, legal, and regulatory considerations and implications for national policy are discussed. Four diagrams, eight tables.

Mixon, W.R. and C.L. Segaser, Energy Conservation Potential of Modular Integrated Utility Systems (MIUS), Intersociety Energy Conversion Engineers Conference, 11th, September 12-17, 1976. Vol. 1, Paper No. 769094, pp 554-558. (New York: American Institute of Chemical Engineers).

Program devoted to development and demonstration of the technical, economic, and institutional advantages of integrating the systems for providing all or several of the utility services for a community is discussed. Possible utility services include electric power, space heating and cooling, potable water, and liquid and solid waste treatment. A MIUS might be sized to accommodate several hundred or a few thousand multifamily dwelling units, nearby single-family housing, and associated commercial facilities. The concept utilizes a relatively small, combined utility plant located on the development site and has the objective of providing utility services in a way that reduces the use of critical natural resources, protects the environment, and minimizes cost. Results related to energy conservation from analytical comparisons of MIUS with conventional utility systems, both serving a 720-unit garden apartment complex, are summarized. Based on a commercially available MIUS model using gas-fueled internal combustion engines to drive electric generators and waste heat from the engines for domestic heating and cooling, MIUS is shown to be significantly more efficient in the use of input fuel energy over a wide range of climate. Energy savings with MIUS range from about 14 percent over the use of conventional heat pumps in San Diego to 50 percent over the use of all-electric apartments with resistance heating in Minneapolis. An overall average of comparisons with five conventional models in five climates amounts to a 30 percent savings of fuel energy with MIUS. The effects on energy use from using solid waste incineration with heat recovery and thermal energy storage in MIUS are also discussed. 8 refs.

Mixon, W.R. and C.L. Segaser, Energy Conservation Potential of Modular Integrated Utility Systems (MIUS), (Energy Research and Development Administration, 1976), 14 pp, GRAI7705, (Contract: W-7405-eng-26; MOnitor: 18).

The Department of Housing and Urban Development is conducting the Modular Integrated Utility System (MIUS) program devoted to development and demonstration of the technical, economic, and institutional advantages of integrating the systems for providing all or several of the utility services for a community. Possible utility services include electric power, space heating and cooling, potable water, and liquid and solid waste treatment. A MIUS might be sized to accommodate several hundred or a few thousand multifamily dwelling units, nearby single-family housing, and associated commercial facilities. The concept utilizes a relatively small combined utility plant located on the development site and has the objective of providing utility services in a way that reduces the use of critical natural resources, protects the environment, and minimizes cost. (ERA citation 02: 000832)
CONF-760906-13

Mixon, W.R., Modular Integrated Utility System (MIUS) as a Potential Influence on Community Development, (Energy Research and Development Administration, 1976), 18 pp, GRAI7626, (Monitor: 18).
This paper summarizes the MIUS concept and the type and extent of potential application. Possible interactions between MIUS and the local community are identified from a comparison of technological and physical characteristics of MIUS with alternative conventional utilities. Possible impacts on land use and patterns of community development are emphasized. (ERA citation 01:021064) CONF-760534-1

Muhlhauser, H. and L.S. Dzung, Combined Power Generation, Paper No. 4, World Energy Conference, 9th, September 1974.

Neal, G.W., "Internal Combustion Engine -- New Roles for Old Performer," Consulting Engineer, Vol. 33, No. 1, (July 1969), pp 88-94.
Mathematical analysis of relevant data for accurate prediction of relative costs based on total energy concept, under arrangement of which engines are used both as source of work for electric generation and as source of heat; how to apply discounted cash flow analysis to determine which type of power will do least expensive work; prime mover heat recovery capabilities.

Nilsson, A.W., "Desalination," ASEA Journal (Sweden), Vol. 45, No. 5, (1972), pp 140-144.

The author presents an economic study made by Stal-Laval Turbin AB of a combined installation for the generation of electrical energy and the production of fresh water. Two alternatives were considered, namely, two units of the Stal-Laval industries' gas turbine type GT35B and two units of the Stal-Laval industrial gas turbine GT120C. Two kinds of evaporator were also included in the study--the multiple effect evaporator and the multistage flash evaporator. Brief descriptions are given of the gas turbines, heat recovery boilers, and evaporators.

Novick, Sheldon, "Electric Power Companies," Environment, Vol. 18, No. 8, (October 1976), p 7.

Difficulties engulfing the electric power industry, including high fuel prices and interest rates, massive rate increases, and nuclear power plant cancellations. (In 1975 cancellations exceeded new orders.) Large commercial and industrial power customers are fighting back at power companies through energy conservation and by non-utility generation. After rising steadily for many years, the load factor of the industry as a whole has begun to decline and, if continued, this would spell disaster for the power industry. Peak load management options are discussed. What is needed is a hiatus of several years in plant orders while demand peaks are reduced and the valley of demand rises slowly. 3 photos.

*Nydick, S.E., et al., A Study of Inplant Electric Power Generation in the Chemical, Petroleum Refining, and Paper and Pulp Industries, Report No. FEA/D-76-321, (June 1976).

The maximum thermodynamic and economic potentials for increased inplant electric power generation by steam turbine, gas turbine, and diesel topping and steam and organic bottoming cycles are determined for the chemical, petroleum refining, and paper and pulp industries with and without government sponsored economic incentives. It was concluded that inplant generation can significantly contribute to U.S. energy conservation strategy. To enhance implementation, a government guaranteed loan program to industry and leasing companies for the financing of inplant electric power generation at current market interest rates is preferred, although it is recognized that a strategy of tax incentives for these investments might be more attractive to certain segments of industry and to utilities. Institutional, legal and regulatory constraints were also assessed. (*Important recent paper.) PB225-659

O'Keefe, W., "In-Plant Electric Generation," Power, Vol. 119, No. 4, (April 1975).

Discusses continuous as well as standing in-plant generation. Outlines various systems available for continuous generation, e.g., steam boiler with steam turbine, condensing turbine, the gas turbine, the reciprocating engine, continued systems or cycles, total energy systems, etc. Discusses feasibility and environmental problems of various systems. Suggests that in standby generation, the engine generator sets are dominant but discusses also other standby sources, e.g., gas turbines, nuclear standby and battery-types standby.

Oplinger, James L., et al., Economic Guidelines for Total Energy Systems, Vol. 80, Part 2, (ASHRAE), Paper No. 2318, (1974), pp 212-223.

The Office of Policy Development and Research of the U.S. Department of Housing and Urban Development called for an analysis of the economic evaluation of total energy with the overall objective being "preparation of guide criteria for the determination, by the private and public sectors of the economy, the applicability of total energy systems for residential development." The purpose of the study described was to identify those situations in which a full technical and economic study is justified. Twelve references.

Orlando, J.A. and D.R. Limaye, Energy Systems for Development, (Pittsburgh: ISA), Annual Conference on Modeling and Simulation, (April 24-26, 1974), pp 673-682.

Description of the nature of the MIUS; reports on analytic investigations of its economics, and describes some of the institutional problems which will impede its widespread use. 4 references.

O'Shea, John, "Perspective: Electricity Too Cheap to Save," Electrical Review, Vol. 197, No. 25, (December 12, 1975), p 772.

Electricity is presently being sold too cheaply to industry, thus reducing any incentive for energy conservation. An integrated energy system producing electricity and heat, applied where appropriate throughout Britain, could result in substantial energy savings. Efficiency of fuel use is

65-80 percent in such a system. Other techniques for energy conservation in utility production and use are described, and costs are presented. The incentive for all these schemes is higher electricity pricing and tax savings for industries applying these schemes. Three diagrams, one table.

O'Shea, J., "Conservation," Engineering, Vol. 216, No. 6, (June 1976), p 411.

In the many cases where factories generate all or part of their electric power and at the same time have a demand for heating or process steam, the use of a combined cycle can show extremely considerable savings. Some of the basic systems -- back-pressure steam, reciprocating engines, and gas turbines -- are reviewed and sample calculations are given for assessing the cost saving. Three diagrams, three photos, one table.

Papamarcos, John, "Combined Cycles and Refined Coal," Power Engineering, Vol. 80, NO. 12, (December 1976), p 34.

For the next several decades, energy conversion efficiencies will continue to be of considerable concern, and the most likely system for providing higher efficiencies in larger power plants is the combined cycle using a clean fuel. The prospect now is low and/or intermediate BTU gas made from coal. The status of gasification programs and fluidized-bed combined cycle, advanced fixed bed gasifier, fluidized-bed boiler, gas cleanup, and the high temperature turbine program awarded by ERDA are described. The first integrated gasification combined cycle in the United States is scheduled to be built at the powerton station of the Commonwealth Edison Company. Five diagrams, three photos, two tables.

Papamarcos, John, "Industrial Power Plants: New Options and Problems," Power Engineering, Vol. 79, No. 7, (July 1975), p 40.

The past growths of the power industry are surveyed over the last 50 years. On-site generation is discussed as having much to offer in increasing plant energy efficiency as it facilitates the use of heat recovery equipment. Prospects for on-site generation, steam cycles, stack gas desulfurization, gas turbines and combined cycles, organic cycles, and integrated systems are detailed. One diagram, one graph, five photos.

Pass, V.H., Lease Financing for Total Energy Systems, Proceedings of Conference on Total Energy 1971, Vol. 1,(Nov. 29 - Dec. 1, 1971),(London: Institute of Fuel).

The author indicates and discusses generally some of the alternative methods of financing the acquisition of an asset, but places particular emphasis on acquiring the use of a total energy system by way of a leasing contract. The two main types of leasing contract are explained. An assessment of three alternative financing methods is made using certain stated assumptions and a statement detailing/comparing the resultant cost of each alternative is given. The author then discusses the 'present value' method of evaluation and shows in graphical form the results achieved using this method. From this analysis is drawn the conclusion that it would be wrong for a company to make its financing decisions on the basis of absolute cost alone, as often what appears at first sight to be the more expensive route may in fact be cheaper in the long run. Finally, some hints for prospective lessees are given, together with a short account of some of the experiences arising from the financing/promoting of total energy systems in the United States. Five references.

Patel, J.G. and J.W. Loeding,"Total Energy Concept Based on Low Btu Gas,"Illinois Institute of Technology Resources Symposium on Efficient Use of Fuels in the Metallurgical Industry, Dec. 9-13, 1974. pp 167-184. (Chicago: Institute of Gas Technology)

A low Btu gasification plant can be designed to supply several forms of energy from a single facility, for example, fuel gas, steam or electric power. These three forms of energy can be produced in a variety of mixes from an extreme of all steam in one case to all electrical power in another case. A low Btu coal gasification plant, therefore, can constitute a total energy system for supplying various energy requirements of a large chemical or metallurgical plant. Four cases are examined to illustrate how the low Btu gasification can be applied in an energy production system.

Payne, F. William and Milton Meckler, "A Marketing 'Blueprint' for the New Energy Era," IEEE Transactions on Industry Applications, Vol. 12, No. 3, (June 1976), p 221.

The market for small electrical generating sets will increase from $782 million in 1975 to $1.8 billion in 1985, representing an average increase of 12.6 percent per year. This market includes internal combustion engines that drive an electric generator mounted on a common base. The rising cost of electricity will spur much of the growth. Two areas of opportunity are explored, generating sets for on-site power generation and for peak load control.

Peters, D., "Exhaust Gases are Set to Work in Total Energy Plant," Engineer (Great Britain), Vol. 232, No. 5999, (January 14, 1971), p 43.

Gas turbine exhaust gases are used both to dry and to convey granular materials in a newly-developed total energy system described by the author.

Peters, C.M.D., Total Energy Systems, (New York: Pergamon Press, Inc., 1976), pp 587-608.

Power generation with heat recovery in external and internal combustion engines is discussed, and efficiencies of various prime movers are compared. Energy resources calculations show savings under load conditions suitable for steam turbine installation. Thirteen references.

Plaskura, W., et al., "Application of an Indirect Heat-Transfer Medium in a Catalytic Sulfuric Acid Plant," International Chemical Engineering, Vol. 14, No. 2, (April 1974), pp 292-297.

An analysis is presented of the economics of using Dowtherm instead of high pressure steam generation for heat recovery in sulfuric acid plants. Specifically, investment and operating costs for three alternatives are given. These are: (1) heat utilization for heating services; (2) heat utilization for electric power generation; (3) heat utilization in a mixed system for heating services and electric power generation. Analysis indicates that the most desirable and economic mode of heat recovery in H/250/4 plants is option number one for plants in the 400-500 ton per day size. For larger plants, option number one or three is preferred. Six refs.

Poradek, J.C., MIUS Wastewater Technology Evaluation, (Houston: National Aeronautics and Space Administration), Report No. NASA-TM-X-58130, JSC-09304, (May 1976), 127 pp.,(Monitor: 18).

A modular integrated utility system wastewater treatment process is described. Research in the field of wastewater treatment is reviewed, treatment processes are specified and evaluated, and recommendations for system use are made. The treatment processes evaluated are in the broad categories of preparatory, primary, secondary, tertiary treatment, physical-chemical processing, dissolved-solids removal, disinfection, sludge processing, and separate systems. Capital, operating, and maintenance costs are estimated and extensive references are given. N76-30120/9ST

Pringle, L.M.,Jr., Technology Evaluation of Control/ Monitoring Systems for MIUS Application, (Houston: National Aeronautics and Space Administration), Report No. NASA-TM-X-58135, JSC-08973, (August 1974), 65 pp, (Monitor: 18).

Potential ways of providing control and monitoring for the Modular Integrated Utility System (MIUS) program are elaborated. Control and monitoring hardware and operational systems are described. The requirements for the MIUS program and the development requirements are discussed. N74-32715/6SL

Redding, Tony E., Application of the Integrated Utilities Concept to Community-Size Developments, (New York: American Society of Mechanical Engineers), Paper No. 749016, (August 1974), pp 493-498. Intersociety Energy Conversion Engineer Conference, 9th, San Francisco, CA.

NASA, along with other government agencies, is conducting a project entitled "MIUS (Modular Integrated Utility Systems)" sponsored by HUD. The MIUS concept combines the utility services of electrical power, heating and cooling, water supply and wastewater treatment, and solid waste management into a single, local plant while conserving energy. The objective of the MIUS project is to demonstrate the technical and economic feasibility of this concept. The results of this study showed that the MIUS concept can be applied to community-size developments with major benefits of energy (fossil-fuels) savings,reduction in solid waste haul-off quantity, and reduction in fresh water demand. Compared to a conventional utilities system, MIUS energy savings were 38 percent, water savings of 17 percent, and solid waste load reduction of 80 percent. Moreover, these benefits are obtainable through the use of commercially-available hardware and with costs competitive with typical conventional utilities.

Reese, T.G. and R.C. Wadle, General Survey of Solid Waste Management, NTIS Report No. N74-27523/1WP, (May 1974).

Potential ways of providing solid waste management for a building complex serviced by a modular integrated utility system were explored. Literature surveys were conducted to investigate both conventional and unusual systems. The merits and disadvantages of the systems most compatible with MIUS are discussed.

Reistad, G.M., et al., "Available Energy and Economic Analyses of Total Energy Systems," Proceedings of the American Power Conference, (April 21-23, 1970), pp 603-611. (Chicago: American Power Conference)

Although giant utilities produce electricity most efficiently, there are still cases where a total energy system with a relatively inefficient power plant is preferable. This paper presents the thermodynamic reasons for the paradox and introduces the engineer to availability concepts. Availability, unlike energy analyses, illustrates and evaluates the thermodynamic losses and, therefore, shows the direction for design improvements. These considerations, combined with the essential economic factors, dictate the minimum cost. Charts are presented to show when total energy with spark-ignition (SI), compression-ignition (CI) or gas turbine (GT) prime movers is economically desirable as a function of fuel and electrical costs, ratio of heat to electrical load, and load factor.

*Resource Planning Associates, Inc., The Potential for Cogeneration Development in Six Major Industries by 1985. Prepared for the Department of Energy Contract #CR-04-60172-00, Cambridge, MA, RPA, (Dec. 1977).

This report examines the cogeneration that can be expected with and without government action in six industries: chemical, petroleum refining, paper and pulp, steel, food processing, and textile. The total potential was reduced considering technical suitability and financial attractiveness. With these reductions the process steam cogeneration in the six industries is expected to achieve only 35% of its 1985 maximum technical potential or 1250 trillion Btu of process steam providing an additional 6,000 Mw of electric power capacity which corresponds to energy savings of 250-420 thousand BPDE which should be largely oil savings. It is estimated that governmental action could stimulate almost as much process steam generation by 1985 as the

*Important recent pages.

baseline development that would normally occur resulting in gross energy savings of 40-140 thousand BPDE depending on the range of federal economic and regulatory measures implemented. Exhibits provide detailed information on the results of this study.

Ritchie, P.J.S.,"Engines Other Than Gas Turbines," Vol. 1, Proceedings of Conference on Total Energy, Nov 29-Dec 1, 1971 Brighton, England. pp 323-334. (London: Institute of Fuel, December 1971).

Development of the medium-speed diesel engine into a "second generation" with brake mean effective pressure (BMEP) of 2.07 10/sup 6/ N/M/sup 2/ (300 lb/in/sup 2/) is reviewed briefly and the problems of developing a "third generation" with predicted outputs of about 2.8 10/sup 6/ N/M/sup 2/ (400 lb/in/sup 2/) BMEP by 1980 are discussed. Electrical generating costs are examined showing that probable performance improvements will give only small reductions and that design must be forced on automation and reliability in order to reduce operating and maintenance overheads. The utilization of rejected heat in total energy systems is shown to be of increasing economic importance. Increased exploitation of natural gas resources could encourage the wider use of gas engines. Possible trends in dual-fuel and spark-ignition systems are examined. The possibility that legislation limiting noise and gaseous emissions could require the reorientation of direct injection combustion chambers in large engines is explained. The Stirling cycle system is reviewed briefly as a possible alternative if limits are set severely low. Three references.

Robb, G.A., Energy Supply and Utilization. Cost Reduction--Conservation, (Montreal: Canadian Pulp and Paper Association), Paper No. T98, (1974).

Recent substantial increases in energy prices, particularly those resulting from collective action by the eleven-nation Organization of Petroleum Exporting Countries (OPEC) has focused attention on energy utilization. Methods of allocating fuel cost to various processes and between process and by-product power are discussed. Techniques for pinpointing cost reduction potential through energy conservation are dealt with as are methods of realizing savings. Total energy systems, recycling of heat, use of groundwood whitewater, more effective use of fuels derived from wood, and greater awareness of incremental energy costs all contribute. Seven references.

Roberts, C.A. and E.J. Perry, Statistical Analysis of Integrated Energy Systems, (Diesel Engineer Users Association), Publication No. 363, (July 1974), pp 1-15.

Characteristics of prime movers are discussed in relation to the demand patterns and the use of demand profiles is described. Examples of cost analyses are given and selection methods discussed. For the purposes of examining security of supply afforded by a given system, a simulation method is described which considers random machine failures and industrial dispute leading to supply interruption. Some examples of simulations are given and these show the variations in security resulting from varying configurations of generating capacity and reliance on the grid system.

Robson, Fred L., et al., "Environmental Impact of Coal-Based Advanced Power Generating Systems," Applied Energy, Vol. 1, No. 2, (April 1975), pp 83-105.

The paper describes an efficient, low-cost power generating system that could utilize high sulfur fuel, designated as the Combined Gas and Steam (COGAS) power system. This advanced power generating scheme has a gas turbine exhausting into a waste heat recovery boiler which raises steam for use in a steam turbogenerator. The factors affecting the performance of combined-cycle power systems and the compromises in the system that result when this system is integrated with a coal gasifier and fuel gas cleanup process are discussed. Seven references.

Rochelle, W.C., et al., Computer Simulation of Thermal and Fluid Systems for MIUS Integration and Subsystems Test (MIST) Laboratory, Intersociety Energy Conversion Engineer Conference, 10th, (Newark, DE) August 18-22, 1975, Paper 759073, pp 461-467. (New York: Institute of Electrical and Electronic Engineers, Catalog No: 75CHO 983-7 TAB).

This paper describes the application of SINDA (systems Improved Numerical Differencing Analyzer) computer program to simulate the operation of the NASA/JSC MIUS Integration and Subsystems Test (MIST) Laboratory. The MIST Laboratory is designed to test the integration capability of the following subsystems of a Modular Integrated Utility System (MIUS): (1) electric power generation; (2) space heating and cooling; (3) solid waste disposal; (4) potable water supply; and (5) waste water treatment. Three references.

Rothenberg, J.H., et al., Integrated Waste Management Systems: On-Site MIUS Applications, (1975), 27 pp, (Monitor: 18).

The Department of HUD is conducting the MIUS program devoted to development and demonstration of the technical, economic, and institutional advantages of integrating the systems for providing all or several of the utility services for a community. Possible utility services include the provision of electric power, space heating and cooling, potable water, fire protection water, and liquid and solid waste treatment and disposal. The objective of the MIUS concept is to provide the desired utility services in a way that is consistent with reduction in the use of critical natural resources, protection of the environment, and minimization of cost. A MIUS might be sized to accommodate several hundred or a few thousand multi-family dwelling units, nearby single-family housing, and associated commercial facilities. It uses a complex of small, on-site combined package plants, each sized to meet demands of the development served. The MIUS is modular in that it can be installed near appropriate users in phase with the actual demands of community development or redevelopment. It uses an integrated systems approach whereby some resource requirements of one service are met by utilizing the effluent of another. For example, heat rejected from electrical generation might be used for residential space heating, and the effluent from liquid waste treatment might be used for fire protection and/or cooling tower water. Each of the major utility subsystems of MIUS are examined with respect to the potential for integration with other subsystems and the advantages resulting from such integration. The conclusions presented represent a current summary of results from evaluation studies of major components and subsystems applicable to MIUS, systems analyses comparing MIUS to conventional utilities serving identical consumer models, and assessments of impacts from application of the MIUS concept. CONF-751216-1

Rothenberg, Jerome H., "Modular Integrated Utility Systems (MIUS) Program," Intersociety Energy Conversion Engineer Conference, 11th, Nevada. Sept. 12-17, 1976. (New York: American Institute of Chemical Engineers), Paper No. 769297, pp 1719-1725.

MIUS is an "extension of a total energy system" and, in addition to performaing energy-heat recovery functions, it also processes solid wastes and liquid wastes and purifies water. In addition to saving energy, MIUS minimizes the adverse environmental impact of utility systems. HUD has organized the program into three phases. Phase I includes evaluation of available technology, development of component and system performance specifications, and detailed technical, economic, and institutional feasibility studies of MIUS concept. At present, HUD-PDR is conducting a demonstration of a total energy system in Jersey City, New Jersey, coupled with an "automatic" pneumatic trash collection system. A MIUS Integration and Systems Test (MIST) facility is being used for the development and verification of MIUS control systems. A coal-fueled MIUS is being developed which will enable the use of coal as a substitute for oil in MIUS energy systems. Phase II of the HUD-MIUS program, which involves the construction and evaluation of a demonstration project in a private sector residential development, was initiated in April 1976. Phase II of the HUD-MIUS program provides for the dissemination of the results and new studies directly oriented to implementation of the concept.

Samuels, G., et al., MIUS Systems Analysis: Initial Comparisons of Modular-Sized Integrated Utility Systems and Conventional Systems, (Energy Research and Development Administration), (June 1976), 266 pp, GRAI7707, (Contract: W-7405-eng-26; Monitor: 18).

This memorandum presents the results of a preliminary study of the economics, fuel requirements and environmental impacts of Modular Integrated Utility Systems (MIUS) and conventional energy systems. The primary analytical effort was placed on determining the costs and energy requirements associated with several methods of supplying electricity, space heating, domestic hot water, and air-conditioning services at sites similar to those in Philadelphia. A total of 11 models are considered. Nine of the models studied are garden apartment complexes; eight have 720 dwelling units, and one has 1080 units. All units have a floor area of 900 feet, except two, and are arranged 12 units to a building. The other two models are 800 feet, except two units in a high-rise apartment building containing 130 dwellings. Gas engines were selected as the prime mover for the total energy systems in the study. The utility grid in the conventional system models is assumed to consist of an equal number of coal-fired and nuclear large base-loaded plants. Emissions of NO/sub x/, SO sub 2, and particulates from the central power station were based on recent federal standards for coal- and oil-fired steam generators with capacities greater than 250 x 10 except 6 Btu per hour. Because no federal standards have been set for CO and hydrocarbons, emissions of these pollutants were estimated from data on four coal-fired steam generators and from data on fuel oil combustion. Data on emissions from gas engines and gas-fired boilers were not readily available at the time of this study; therefore, for this initial comparison of the effects on air quality, the prime movers of electrical generators of the first model were assumed to be diesel engines and the fuel for on-site boilers was assumed to be NO. 2 fuel oil. Additional information is provided in 11 appendixes. (ERA citation 02:005168) ORNL/HUD/MIUS-6

Samuels, G. and W.J. Boegly Jr., MIUS Technology Evaluation: Preliminary Subsystem Recommendations for Near-Term Concepts, (Energy Research and Development Administration, May 1976), 60 pp, GRAI7626, (Contract: W-7405-eng-26; Monitor: 18).

The MIUS program, under the overall direction of HUD, is directed towards the development, demonstration, evaluation, and ultimate widespread application of a new option for providing utility services to communities. The MIUS concept is to provide developing communities with energy, water, and sanitary services from a combined on-site utility plant. A total system approach is used to balance the requirements for environmental quality, conservation of resources, and low total cost. The analysis reported herein is one of a series of evaluations of utility technologies applicable to MIUS. A few examples of what major components and subsystems might be incorporated into a typical MIUS are described and discussed, together with possible methods of integration. Maximum effort has been made to include all possible community services: power generation, space heating and cooling, domestic hot water, potable water, fire protection water, and liquid and solid waste treatment and disposal. In some of the examples cited, complete thermal integration has not been obtained, and in certain cases integration of the liquid waste subsystem is merely the use of treated effluents for fire protection or cooling tower makeup. The final selection of the various subsystems used in a MIUS will depend on the climate, population density, total population, consumer mix and demands, local geology and topography, and local resources such as water, natural gas, oil,

and land. Schematic diagrams are provided with several MIUS configurations; however, it must be pointed out that these are not necessarily the "optimum" or "best" configurations. The main purpose of this report is to illustrate how MIUS can be assembled using available technologies and some of the alternate possibilities. (ERA citation 01:019448)
ORNL/HUD/MIUS-17

Schaefgen, John R., Integrating Community Utilities for Resource Conservation, (Washington, D.C.: Nat'l Bureau of Standards, 1976), 9 pp, GRAI7624, (Monitor: 18).

A MIUS provides the utility services of electrical power, space heating and cooling, potable water heating, solid waste processing, sanitary sewage treatment, potable water treatment, and the site distribution, collection, and disposal functions attendant thereto. MIUS facilities are constructed on a community scale to achieve the maximum symbiotic effects from process integration. The MIUS approach to utility services can conserve energy, reduce environmental degradation, and permit greater flexibility in intensive land development at a competitive cost. This report briefly discusses the characteristics and benefits of MIUS facilities. PB-256 898/8ST

Schwieger, R.G., "Future Brightens for Combined-Cycle Plants," Power, Vol. 115, No. 10, (October 1971), pp 105-109.

Combined-cycle plants, wherein gas turbines, heat-recovery boilers and steam turbines are matched to provide an optimum system have been pre-engineered with packaged components tailored to meet utility mid-load generation demands.

Sebald, Joseph F., Report on Site and Design Temperature Related Economics of Nuclear Power Plants with Evaporative and Non-Evaporative Cooling Tower Systems, (Springfield, VA: National Technical Information Service, January 1976). NSA3312

Comprehensive study of LWR and HTGR plants to include a wide range of plant design temperatures, develop cost elements suitable for the adjustment of differences in site, time, and location, and to develop cost sensitivity criteria such that the mass of cost data produced may be extended to include plants sited most anywhere in the United States, with acceptable accuracy is presented. The Lloyd site with evaporative, and the West Valley site with non-evaporative cooling are described and engineering costs are evaluated. Unit cost comparisons between LWR and HTGR plants and comparison of equivalent plants of both sites are presented. COO-2392-1

Segaser, C.L. and W.R. Mixon, MIUS Systems Analysis: The Effects of Thermal Energy Storage and Solid Waste Incineration Options on MIUS Cost and Fuel Consumption, (Energy Research and Development Administration, July 1976), 64 pp, GRAI7705, (Contract: W-7405-eng-26; Monitor: 18)

Water is the preferred fluid for the conveyance of thermal energy between a MIUS district heating and cooling plant and the consumer. Because of its abundance, low cost, and unique thermal properties, water was also chosen as the thermal energy storage (TES) medium for MIUS in preference to one of the available latent-heat phase-change materials. This report presents the results of a parametric system analysis using the ORNL 1973 MIUS hourly comparative energy computer program to evaluate the effects of a sensible heat thermal energy storage device on fuel consumption in a hypothetical MIUS model. An economic evaluation is included for cases with and without solid waste incineration and heat recovery over a range of storage capacities and fuel costs. The analysis indicates that a significant savings in fuel consumption can accrue from the storage of excess waste heat for subsequent use during periods when the available heat from the engine-generators is not adequate to satisfy the demand. Use of the heat energy available from the incineration of solid wastes can result in a fuel savings of up to five percent per year if a water-storage vessel of 125,000 gallon capacity is used. A vessel of this size is capable of storing 4.0 x 10 exp 7 Btu of heat energy from an initial temperature of 160 exp 0 F to a final temperature of 200 exp 0 F. The breakeven cost of the fuel (natural gas), at which the value of the fuel saved equals the estimated yearly owning cost of the thermal energy storage facility, was shown to be of the order of $2.25 per thousand cubic feet. The economic analysis was extended to compare the costs of various methods of solid waste incineration for the MIUS and consumer models considered. The options included combinations of cases with and without heat recovery, thermal storage, and supplemental fuel consumption. The incremental fuel cost or credit for each option was based on the difference in total fuel consumption from a MIUS model without incineration. (ERA citation 02:000833)

Segaser, C.L., MIUS Technology Evaluation: Thermal Energy Storage Materials and Devices, August 1975, (November 1975), 65 pp, (Contract: W-7405-eng-26; Monitor: 18).

The generation of electricity by an on-site power plant in a modular integrated utility system (MIUS) results in a considerable amount of waste heat energy being released from the prime mover coolant system and exhaust gases, which can be effectively utilized for domestic space and water heating and, with less efficiency, for domestic air-conditioning by use of an absorption chiller. If the amount of waste heat that can be recovered at any instant does not closely match the heat demand, excess heat must be rejected either to the environment or stored until it is required; a heat deficiency must be made up from storage or an auxiliary boiler. An evaluation is presented of the technologies relevant to the design of thermal energy storage (TES) devices and the selection of thermal energy storage materials for application in a MIUS. Several thermal energy storage materials are commercially available in large quantities, are relatively inexpensive, and have properties which make them suitable for use in a MIUS. The techniques for thermal energy storage may involve the sensible heat of water or one of the available phase-change materials. The conclusion is that water is a logical choice for thermal energy storage in a MIUS. This choice is based partly on the fact that water is a low cost material. Also, because water is the most likely thermal energy conveyance medium between a MIUS district heating and cooling plant and the point of use, the thermal energy storage containment unit can be a simple large district storage tank that does not require an intermediate heat exchanger. ORNL-HUD-MIUS-23

Sharygin, V.S., "Choice of Structure of Peaking Electric Power Plant for the Total Energy System in the North West," Problems of Power Development: North West USSR, (1971), pp 195-203. Published by Zinatne Riga

The technico-economic characteristics of gas turbine and pumped storage hydroelectric plant are summarized. A method is proposed for optimizing the choice of distribution of peaking capacity in the North West USSR system, and results of a calculation using the method are presented and discussed.

Shields, Vernon and Joseph M. King, Jr.,"Application of Fuel Cells with Heat Recovery for Integrated Utility Systems," Intersociety Energy Conversion Engineer Conference, 10th, Newark, DE, Aug 18-22, 1975. Paper No. 759044, pp 278-281. (New York: Institute of Electrical and Electronic Engineers, Catalog No. 75CHO 983-7 TAB)

This paper presents the results of a study of fuel cell power plants with heat recovery for use in an integrated utility system. Such a design provides for a low pollution, noise-free, highly efficient integrated utility. Use of the waste heat from the fuel cell power plant in an integrated utility system for the village center complex of a new community results in a reduction in resource consumption of 42 percent, compared to conventional methods. In addition, the system has the potential of operating on fuels produced from waste materials. 2 refs.

Shirazi, Mostafa A., Thermoelectric Generators Powered by Thermal Waste from Electric Power Plants, (October 1970), 30 pp, GRAI7210, (Project No: EPA-WQO-16130; Monitor: EPA-WQO-16130-10/70).

The feasibility of recovering electricity from the waste heat of electric power plants was assessed. Sources considered were stack flue gas, gas turbine exhaust, and condensing steam. Typical 1600 mw fossil fuel steam plants and gas turbine plants were used as examples. Flat plate heat exchangers were designed with thermoelectric couples arranged in series within the plates. Heat flux, conversion efficiencies, and flow friction losses were calculated. Except for the condenser application, the friction losses are several times the thermoelectric power generated. Under favorable conditions, 3 to 9 mw is obtainable from the thermoelectric condensers. The high material cost, however, precludes all such applications today. PB-207 870

Simon, M., "Total Energy -- One Way to Save Energy," M.A.N. Resource Engineering Manufacturing (Germany), No. 6, (April 1975).

Total energy is the optimum utilization of primary energy when it is converted into mechanical-electrical energy, while exploiting the heat of exhaust gas and cooling water. The author discusses the use of internal combustion engines as part of total energy systems. The need for such systems, their applications, and advantages are then considered.

Smith,A.J., "Gas Turbine Retrofit Pays Dividends," Electrical World, Vol. 181, No. 2, (January 1974) pp 62-63.

Heat recovery combined cycle (HRCC) systems have been proven as a means of increasing the electric power generation in gas turbine installations by recovering the energy in the hot air exhausts energy that is otherwise lost. HRCC is accomplished by installing an unfired heat recovery boiler in the exhaust of an open cycle gas turbine, and adding a condensing steam turbine for power generation. Compared to other alternatives for energy conservation or to new generating capacity, the addition of the HRCC to open cycle peaking gas turbine installations has several significant advantages.

Smith, E. and W. Aicher, "Heat Recovery Boilers for Total Energy Systems," Total Energy Conference, Vol. 1, Brighton, England,(Nov. 29-Dec. 1, 1971). Paper No. 11, pp 185-209. (London: Institute of Fuel).

The addition of a heat recovery system to a gas turbine plant is an essential requirement for a total energy scheme. One form of recovering the heat rejected by the turbine is to generate steam in waste-heat boilers. Such heat recovery is limited by the turbine exhaust gas weight and temperature, and the pressure of the steam required. The relationship of these factors is discussed. Additional steam can be generated by raising the exhaust gas temperature by direct injection of liquid/gaseous fuel. Gas turbine design parameters which influence boiler design are discussed and illustrated by basic turbine design equations and graphs. A review is also presented of the costs for a gas turbine driven compressor total energy system compared with a motor-driven compressor system.

Snell, Jack E. and Paul R. Achenbach, Total Energy Systems: A Review of Recent NBS Activities, (Washington, D.C.: National Bureau of Standards) Final Report, (June 1972), 33 pp, GRAI7505, (Monitor: 18).

This paper briefly reviews and reports on the status of several NBS activities in the area of total energy systems as they relate to the theme of the Drexel Symposium on Effective Energy Utilization. A recent review of the state-of-the-art in total energy systems applications indicated the need for good, concise, operating and performance data on total energy systems to serve as a basis for more effective utilization of this concept to save energy.

HUD requested NBS to assist in the development of a total energy system on the Breakthrough housing site at Jersey City and to conduct a full-scale field study with appropriate instrumentation to obtain engineering, cost, and maintenance data. In a related effort, NBS, AEC, EPA, and NASA are supporting HUD in a new effort entitled the MIUS program. The premise of this three-phase effort is to demonstrate the potential energy, resource, and cost economics of combining energy-generating facilities and waste disposal facilities on a modular basis to keep pace with changing patterns and rates of urban growth.

COM-74-51130/4ST

Stetkar, J.W. and M.W. Golay, Design of an Optimal Total Energy System for a Large Military Installation, (American Society of Mechanical Engineers), Paper No. 76-WA/HT-46, (December 1976), 9 pp.

The use of a recently developed computer code for simulating the thermal and electrical energy demand behavior of a multi-consumer-type community and for modeling the community's thermal energy utility system is discussed in conjunction with design studies being performed for a proposed total energy system at Fort Bragg, North Carolina. Optimized nuclear HTGR-powered and fossil CGGT-powered utility systems are shown to be capable of meeting the base's annual total energy demands at a thermal-to-electrical energy supply ratio of approximately 75 percent: 25 percent, with the fossil option affording the minimum total system cost-over-life in 1985 dollars. Nine references.

Stetkar, John W., et al., Design of a Nuclear-Powered Total Energy System for Fort Bragg, North Carolina, Final Report. (May 31, 1976) 306 pp, GRAI7701, (Contract: DAAK02-74-C-0308; Monitor: USAFESA-RT-2011).

A total energy system (TES) is designed to supply the thermal and electrical energy requirements of Fort Bragg, North Carolina for a period of 30 years, with startup scheduled for early 1985. Considered for use as the central station power plant for this system are a combined coal gasification, fossil-fired gas turbine (CGGT) power plant and a direct Brayton cycle high temperature gas cooled reactor, helium gas, turbine (HTGR/GT) power plant. Several utility system configurations affording different thermal/electrical energy demand ratios are studied for each supply option. With the primary system optimization criterion being the choice of the total energy system providing minimum total energy costs over the system lifetime, it is found that the optimal thermal/electrical load split for each supply option occurs at approximately 75 percent of the base's total energy demands supplied thermally. Within the limits of the unit-cost assumptions made and for the range of cases studied, it is found that the present-worth total cost of the optimized HTGR/GT system (in 1985 dollars) is $245.7 million and the corresponding optimal system cost for the fossil CGGT alternative is $181.7 million. Further studies are recommended to investigate the sensitivity of this 26 percent cost differential to variations in the mode of power plant operation and to design modifications in the thermal energy distribution piping network. AD-AO31 744/6ST

Stetkar, J.W. and M.W. Golay, "Simulation of the Behavior of a Brayton-Cycle High-Powered Thermal Utility System for Total Energy Application," American Nuclear Society, Vol. 22, (November 1975). pp 601-602.

The use of a newly developed computer modeling program to aid in the design and performance analysis of the thermal distribution portion of a total energy system is described. The program provides the capabilities for simulating both the steady state and time dependent behavior of the energy consumers and the supply network serving them. Included in the model are detailed descriptions of the transient energy and momentum transfer within the high temperature water supply pipes, heat exchangers, and circulation pumps in the distribution system, the energy source at the HTGR/GR power plant precooler and a central high temperature water thermal energy storage reservoir for system load smoothing. 4 refs.

Stetkar, John W. and Michael W. Golay, TDIST, A Program for Community Energy Demand Analysis and Total Energy System Response Simulation. User's Manual, (Massachusetts Institute of Technology, Department of Nuclear Engineering), Final Report, (August 1976), 296 pp, GRAI7705, (Contract: DAAK02-74-C-0308; Monitor: USAFESA-RT-2024).

The thermal distribution system simulation code (TDIST) is a computer program developed at MIT to aid in the design and dynamic performance analysis of a large integrated total energy system supplying thermal and electrical energy to a multi-consumer-type metropolitan area. Included in this manual are derivation of the detailed consumer energy demand models and of the thermal utility system (TUS) component models developed in each of the system analysis subroutines. Comprehensive modelling scenarios are presented to illustrate how these program components are combined to yield a range of steady-state and dynamic system analysis options. Program execution information includes descriptions of the three primary modes of program operation, complete documentation of input requirements and data card formats, examples of available data output options, and a listing of the code. Sample data is included for the execution of a 24-hour dynamic simulation of a total energy system serving a community of 50,000. AD-A033 301/3ST

Stocks, W.J.R., "Gas Turbine Offers Short Cut to Energy Economy," Energy International, Vol. 11, No.4, (April 1974).

Industrial gas turbines have found widespread application in total energy systems associated with building services or industrial processes. The high thermal efficiencies that are obtainable compel deeper consideration of such schemes as a way of conserving fossil fuel reserves.

Stocks, W.J.R., "Gas Turbines for Total Energy," Electrical Review, Vol. 185, No. 20, (November 1969).

Reliable continuous operation with low routine maintenance requirements make gas turbine generator sets attractive propositions for many total energy schemes, as discussed in this paper.

Stocks, W.J.R., Heat Recovery Systems--General Review Vol. 1, Total Energy Conference, Paper No. 10, (December 1971), pp 169-184.

A general survey is presented of the applications of various types of prime movers in total energy systems, the forms of heat recovery possible as well as the effect of the latter upon the types of prime mover to be selected, and upon the earning capacity of the installation. A simple guide is given to the preliminary assessment of the feasibility of a total energy system, and various typical applications are described.

Strauss, S.D., "Heat Recovery Takes a Fresh Turn," Power, Vol.120, No. 1, (January 1976), pp 19-23.

Revamping of the energy system serving ten institutions adds electric generation and waste recycle to steam and chilled-water supply. Results are marked savings in cost and fuel. Reduced noise and pollution with service to spare for a housing development.

Stromberg, R.P., *A Solar Thermal Electric Power Overview*, (New York: Institute of Electrical and Electronic Engineers, April 1976).
An aggressive program to establish the feasibility of solar energy for producing electricity is being sponsored by the Energy Research and Development Administration. Two major ideas are being pursued. Central power plants, remotely located, would replace conventionally fueled plants. Total energy systems, closer to the developed areas of cities would produce electricity and utilize the large amounts of thermal energy otherwise lost. Economics of solar systems look marginal in contrast to current fuel prices. It will be necessary to change incentives in order to accelerate the transfer from fossil fuels to alternative sources of energy. One ref.

Strong, R.E. and J.A. Sykes Jr., *Pace-SGP Oil Gasification Combined Cycle for Power Generation*, Vol. 36, (Chicago: Illinois Institute of Technology), (May 1974).
The system consists of a combined-cycle generating plant fueled by clean Btu gas produced by partial oxidation of available high-sulfur residual fuel. Critical problems of control, operation, and environmental impact have been carefully considered to determine that the plant will prove to be a reliable source of electric power.

Sutton, K.W., "A Consultant's Approach to Total Energy," *Electronics and Power* (Great Britain), Vol. 21, No. 10, (May 1975).
This article focuses attention on the possible use of total energy systems as a means of reducing overall energy consumption of commercial and industrial users, and indicates some of the considerations involved when initiating an analysis of a total energy system.

Taylor, F.M.H., "Total Energy for Cost Reduction," *British Chemical Engineering*, Vol. 14, No. 7,(July 1969),pp 941-943.
Industrial and commercial organizations pay between 1.25 and 1.8 pence per kilowatt hour for electricity from the Area Boards in Britain. On the other hand, there are a number of installations of on-site power generation where the overall cost of

power is 0.5 pence per unit lower, thus showing a saving of 30 percent. Because on-site generation has frequently been linked with recovery of what was previously rejected energy in the form of engine jacket cooling water and high temperature exhaust gases, the term "total energy" has been introduced to embrace a wide variety of systems where a primary source of energy, such as oil or natural gas, is employed to provide the comprehensive energy requirements in the form of light, heating, cooling, air-conditioning, drying, process heat and power for a works or for a commercial or public building. It thus includes schemes for electricity generation, waste heat recovery and direct power drive of mechanical equipment directly coupled to a prime mover.

Taylor, J.C.B., "Power and Energy from Gas Prime Movers," Plant Engineering, (London), Vol. 18, No.9, (September 1974), pp 17-25.

The extraction of energy from the basic source may be only a small part of the overall efficiency. For example, a large modern boiler may extract nearly 90 percent of the energy from the fuel and convert it to steam, but if only 40 percent of the energy in the steam can be utilized, the overall efficiency would be less than 30 percent. However, efficiency can be greatly increased by heat recovery for process heating. Electricity and process gas production, using gas turbines, are compared with alternate methods.

Taylor, J.R., 120-mw Total Energy Plant at Grenoble, Vol. 1, (London: Institute of Fuel), Paper No. 1, (December 1971), pp 302-311.

Description of the total energy plant at Pont-de-Claix where eight unitized packaged gas turbines with exhaust heat-recovery boilers produce 120-mw of electric power together with some 370 tons of steam an hour. As the gas turbine fuel, a special wax distillate fuel has been adopted. The characteristics of this fuel are given together with notes on handling techniques and its good performance as a gas turbine fuel. Some comparisons are made of the probable cost of purchasing large supplies of industrial power from a public supply system with self-generation and on-site steam raising.

Tintori, J., et al., Fuel for Total Energy, (American Society of Mechanical Engineers), Paper No. 71-GT-55, (April 1971), 7 pp.

This paper reviews the uses of gas turbines in total energy systems, describes the actual fuel adopted, and gives a comprehensive description of the particular plant installed. Some notes on the plant performance up to the time of the submission of the paper are also included.

Tobin, P.M. and C.E. Myers, *Integrated Waste Management Systems*, (American Society of Mechanical Engineers), Paper No. 74-ENAs-47, August 1974, 12 pp

Reviews the past work on Alaska Village Demonstration Projects, siting of public and private utilities, and new solid waste collection techniques. The MIUS under joint sponsorship with HUD and other new concepts including use of solid waste as a supplement fuel source, recovery of methane from organic material, and pyrolyzation of organic wastes are discussed. Finally, reports on several communities which plan to use integrated waste management systems approach are presented. To date the approach of examining pollution control through the total waste management concept appears to be practical and has been receiving enthusiastic acceptance. 19 references.

Tucker, A.C.N. and A.F. Williams, *Various Fuels Used in Total Energy Systems*, Vol. 1, (London: Institute of Fuel), December 1971.

The authors describe commercial and factory total energy systems from the standpoint of the oil and natural gas suppliers. The properties, supply, and storage of fuels are discussed in relation to the operational requirements of the preferred prime movers (diesel/gas engines and gas turbines). Emphasis is laid on the clean, easily handled fuels ranging from natural gas through liquefied petroleum gases, naphtha and gas oil. It is concluded that reciprocating enegines are best suited to office-block type developments with high electrical and low heat loads. Gas turbines in the 5-10-mw range appear attractive for factory estates with high steam/hot water demands, particularly if ash and sulphur-free fuel permits direct use of exhaust gases for process heat.

Valentine, A.C., "Private Power Generation," *APE Engineering* (Great Britain), No. 18, April 1975, pp 2-7.

Discusses various situations where private electrical power generating plant can be considered and outlines several possible generating plant systems, base-load plus public supply standing, base-load supplemented by the public supply, use of process waste, incineration of domestic refuse, peak-load lopping in parallel or not in parallel with public supply, and standby to public supply system. Surveys the basic prime movers for various systems, e.g. diesel engine, dual-fuel diesel engine, steam turbine, and gas turbine.

Wakamatsu, S. and S. Koyama, "Researches on Total Energy System. Preliminary Study on Energy System Model," Bulletin of Electrotechnical Lab (Japan), Vol. 39, No. 10, (1975).

A total energy system is defined as a nation-wide overall energy system consisting of a great number of elements which are closely related to each other and with surrounding systems. For the purpose of estimating future energy systems, finding desirable system configurations, and furnishing guides to system shifts is discussed,a method for evaluating large and complex future energy systems. Some criteria for such systesm have been clarified. Systematic procedures to survey and analyze the basic data, construct an energy system model for quantitative analysis and evaluate the energy system have been developed. 48 references.

Wakefield, R.A. and M.J. Damborg, "Total Energy System Long-Term Feasibility," Modeling and Simulation, Vol. 6,(1975),pp 33-38.

In light of the growing scarcity of basic fuels and environmental concern about the effects of thermal pollution, it is increasingly important to consider the feasibility of thermal power plants that are designed to supply both electricity and useful heat to consumers and to reduce the quantity of waste heat rejected to the environment. The research report is a first attempt at analyzing such feasibility. 8 references.

Weaver, W.J.,Jr., "Total Energy Systems for Enclosed Shopping Malls," Actual Specifying Engineer, Vol.29, No. 5, (May 1973), p 87.

This article explains why total energy systems may be the best choice for powering shopping centers of more than 500,000 square feet. It explains how the systems reduce power costs, make more efficient use of fuel, and cause less pollution.

Wegmann, E., "Energieverbundsystem in Einem Sportzentrum--Integrated Energy System in a Sports Center," Elektrowaerme International (Germany), Vol. 32, No. A3, (May 1974), pp A147-A151.

The sports center described is planned to incorporate an indoor swimming pool, a triple gym, sport grounds, and a skating rink. An attempt is made to show how the problem of costs and that of the burden on the environment as a result of emissions and a high release of heat can be solved with the means available.

White, Laurence J., "Nuclear Heat Seeks Process Applications," Chemical Engineering, Vol. 81, No. 24, (November 1974), p 78.

In 1980, process steam is to flow through twin 48-inch pipelines that will span the two miles between a consumer power company plant and Dow Chemical's Midland Michigan chemical complex. By the middle 1980s, high temperature gas cooled reactors will begin supplying heat for such tasks as coal gasification, water splitting, and steel making. Prospects for applications of nuclear-generated heat are reviewed. One graph.

Willenbrock, F. Kar, "Energy Strategies for Health Care Institutions," Dimensions - NBS, Vol. 59, No.10, (October 1975), p 222.

About one-third of U.S. energy consumption provides building services. Energy use in a building is determined by design, construction, and maintenance. Energy conservation in heath care institutions is discussed. Four projects, including a Tennesse hospital, are outlined. The modular integrated utility system concept and its application to medical complexes are explored. Two photos.

Wilson, W.B. and J.M. Kovacik, Turbine Systems to Reduce Petroleum Refining and Petrochemical Plant Energy Costs, (American Society of Mechanical Engineers), Paper No. 76-Pet-62, (September 1976), 11 pp.

Turbines integrated into the petroleum refinery and petrochemical plant's process heat system can generate power (kw and/or hp) with less than half the fuel required by a separate plant generating and transmitting electric power to the industrial. By use of typical examples, this paper discusses the performance and economics of different energy supply system alternatives and the role turbines can play in providing a reliable power supply system with the operating flexibility needed to meet the process plant's varying requirements for kilowatts, horsepower, and process heat.

Yuengling, D., et al., Energy Recovery System for Ships, Final Report, (December 1976), 183 pp, GRAI7707, (Contract: MA-6-38010; Monitor: MA-RD-930 77014).

The volume reports the results of a feasibility investigation of various heat recovery concepts which may be incorporated in an existing marine steam plant. The use of differnet refrigerant fluids as well as different types of energy recovery units is exmained and a plan developed for prototype development, installation, and test. PB-262 124/1ST